D0329081

COMBUSTION

The Formation and Emission
of Trace Species

by

JOHN B. EDWARDS

Adjunct Professor

Department of Chemical Engineering

University of Detroit

466441

MITCHELL MEMORIAL LIBRARY
MISSISSIPPI STATE UNIVERSITY
MISSISSIPPI STATE, MISSISSIPPI 39762

ann arbor science PUBLISHERS INC.
POST OFFICE BOX 1425 • ANN ARBOR, MICH. 48106

Copyright © 1974 by Ann Arbor Science Publishers, Inc.
P.O. Box 1425, Ann Arbor, Michigan 48106

Library of Congress Catalog Card No. 73-93952
ISBN 0-250-40054-5

All Rights Reserved
Manufactured in the United States of America

TO MY MOTHER
for her gifts of
life and of love

Preface

Combustion is a phenomenon familiar to virtually every man, woman and child. Each of us depends upon various combustion processes to generate the electric power we use to heat our homes and to power our automobiles. Only recently have we realized that the multitude of trace species produced by all of our beloved combustion processes may also have an impact upon us and upon our environment. It was in the midst of this revelation and in the emotion accompanying the Earth Days of 1969 that I was called upon to develop two courses, one at the undergraduate and the other at the graduate level, dealing with the air pollution aspects of combustion processes.

Student interest was high and the response was overwhelming. At the same time the task was monumental for I was soon to discover that the available material was fragmentary. Bits and pieces were scattered here and there in the journals. Much of the information bordered on the frontiers of knowledge, and sometimes it was difficult to separate fact from conjecture. Credit must be given to the students who labored through my lectures in those first courses I taught. For many, that was the first course they took that lacked the ordered presentation of thoughts that a textbook helps provide.

My own training is in chemical engineering and my heavy emphasis on the chemical aspects of the subject will no doubt come through to the reader. I must thank many of my nonchemically oriented students for their suggestions. Hopefully these suggestions have resulted in a book that is more readable to those who have only an elementary knowledge of chemistry. I also hope the approach used in this book will emphasize the importance of chemistry in understanding the formation of trace combustion products and will encourage the reader to improve his understanding of chemistry if it is inadequate.

The reader must be cautioned about one thing. This book deals primarily with the mechanisms by which trace species are formed in and emitted from combustion processes. Thus it emphasizes only one facet of the broader subject of combustion. Even a thorough understanding of the material presented herein is by no stretch of the imagination sufficient to make one an expert on combustion. There are many other important aspects of the subject that must also be understood.

Finally I would like to emphasize that though this book was developed in an academic environment it is intended to serve more purposes than just being a textbook. The subject matter is of contemporary interest to many persons with diverse academic training but with a common professional interest, the emission of air pollutants by combustion processes. For these individuals this book can serve as a reference text.

Special thanks are due to many who assisted in the development of this book. I must thank my wife Dixie for her patience. I must also thank Dr. Frank E. Ammerman who painstakingly read the original manuscript and made innumerable suggestions that have been incorporated in the present book. Credit must be given to Dr. Charles F. Reusch, of the Bureau of Power, Federal Power Commission and to Dr. Robert Barry of Detroit Edison for their valuable contributions and suggestions. Many other of my colleagues deserve mention for their encouragement. Among them are Dr. Clayton Lewis (now retired) and Dr. D. Maxwell Teague of the Chrysler Corporation and Dr. Leon Kowalczyk (now retired) of the University of Detroit.

I must also thank Garnett Ross, Bryan Suits, Peyton Suits and Keith Mann for their valuable suggestions and assistance with the preparation of the figures.

Finally I would like to acknowledge the contributions of a number of other persons who by their friendship and encouragement have made my own environment more pleasant and the task of authorship less of a chore. Among these are Brad and Mark Dreisbach, Todd Holtz, Todd and Tim Johnson, Steve Hertler, Paul and Jeff Yourick, Robert Foster, Peter Harris and Rick DeHart. It is my hope that in some way this book will contribute to making their environments a bit brighter also.

April 1974
John B. Edwards
Ann Arbor, Michigan

TABLE OF CONTENTS

Chapter Four—Diffusion Flames   123

Chapter 5—Applications   167

LIST OF TABLES

LIST OF FIGURES

INTRODUCTION

1.1 OVERVIEW

A multitude of combustion processes are utilized in contemporary societies. Table I illustrates the diversity both of commonplace combustion processes and of the substances used as fuel.

The major focus in this text will be on the mechanisms of formation and emission of trace species from combustion processes. Before becoming embroiled in detailed technical discussion it is worthwhile to consider the broader aspects of the subject just to give the reader an idea of the relevance of the subject matter.

Combustion processes emit a variety of trace species which under certain conditions become air pollutants. That is, their presence in the air at concentrations above a certain level is undesirable. This is likely to occur in urban areas where the densities of people and of the combustion processes associated with mobility, domestic and industrial activities tend to be large.

Smoke, or finely divided particulate matter dispersed in air, is perhaps the most visible and well-recognized sign that combustion processes do emit trace substances. Up until quite recently smoke emanating from stacks was of no great concern to the populace. Indeed for some it was a good sign. During war years posters showing factories billowing forth plumes of smoke were a constant reminder of the relationship between production and national security.

Today attitudes are rapidly changing. Smoke and even stacks with no visible smoke are viewed with suspicion. They are reminders of environmental degradation and waste. It has been recognized that although smoke is a visible sign of atmospheric pollution, combustion processes emit many invisible pollutants. At one time carbon monoxide was the only widely recognized invisible pollutant. Today the names of many others, including nitrogen oxides or NO_x, sulfur oxides and hydrocarbons, are becoming household words.

The reasons for this change in attitude are numerous and complex. The general public has become better informed about environmental

Table I

COMBUSTION PROCESSES & FUELS

Source	Typical Fuel
Mobile	
Automobiles, motorcycles, snowmobiles, lawn mowers, chain saws, portable generators	Gasoline, LPG
Trucks, buses, locomotives, vessels, heavy construction equipment	Diesel
Aircraft	Kerosene, JP-4
Missiles	Solid and liquid propellants
Stationary	
Electric power generation	Natural gas, oil, coal
Space heating (direct fuel and catalytic)	Natural gas, LPG, fuel oil, coal
Waste disposal Open burning—dumps, fields (agricultural), timbers and large objects, leaves Incineration—home, municipal and industrial	Solid and liquid waste materials including wood, paper, rubble, plastics, oils, vegetation, animal wastes, sewage, sludge
Industrial Processes	
Metallurgical roasting; refinery (flares, etc.) and process energy	Coal and misc. organic chemicals
Afterburners for pollution abatement	Natural gas
Miscellaneous	
Cooking	LGP, LNG, charcoal, wood
Smoking	Tobacco
Agricultural smudge pots	Oil
Uncontrolled fires forest, buildings, waste piles, storage piles, materials in shipment	Wood, coal tailings Combustible chemicals
Explosions	Propellants

matters, including the relationship between air pollution and health. The mass media, of course, have played an important role in awakening public awareness. So have other institutions. Courses about the environment have been introduced in our educational institutions from the elementary to the university level. However, the reasons for this change in attitude are more fundamental and more deeply rooted in technology and in society. Recently rapid technological advances have taken place in a number of seemingly unrelated fields, greatly facilitating our understanding of combustion processes, the species they emit and their interactions with and impact on our environment. Possibly the most notable example of this is the developments that have taken place in electronics. One consequence is that analytical techniques have evolved rapidly. A decade ago measurements of trace species were largely accomplished by using either wet chemical methods, which were time consuming, cumbersome and expensive because they required highly skilled personnel, or by the use of simpler but often rather crude and inaccurate shortcuts. Today instrumental techniques incorporating solid state electronic and other devices such as lasers are available. Many of these instruments are more sensitive, specific and accurate than their predecessors. In many cases they lend themselves to making continuous real time measurements. Large amounts of data on complex systems can be accumulated rapidly and rather inexpensively by relatively unskilled operators.

Recent advances that have been made in the field of electronics have led to the development of modern digital computers that make it feasible to construct numerical models of complex processes such as combustion and even of the environment. The predictions of these models have added to our understanding of physical and chemical processes, notably in the area of kinetics. Advances in our ability to ·model these systems have been so rapid indeed that in many cases they have outpaced our ability to supply the necessary input data.

As a result of these advances in instrumentation and modeling, problem and potential problem areas have been uncovered that may have been present for decades but were previously unrecognized. However, it should be recognized that not all of the current concerns about trace species emitted by combustion processes revolve about subjects that were simply dormant before. Some of the problems are uniquely today's problems. A trend that has intensified during the twentieth century is urbanization. Large metropolitan areas have and are continuing to form. Their development is particularly rapid along the major traffic arteries connecting adjacent metropolitan areas, and result in population centers called megalopolises or conurbations which sometimes stretch for hundreds of miles. The demands for light, heat, electrical power and waste disposal by the millions of residents in each of these areas result in rates of local emissions of trace species exceeding the capacity of the atmosphere to disperse or otherwise assimilate them. If their concentrations in the atmosphere rise, undesirable effects may result.

A detailed discussion of the many possible effects of trace species that may be emitted to the atmosphere is beyond the scope of this discussion. These include health effects for humans and animals, plant and material damage, contamination of soil and water resources and perhaps even alterations in the local and global climate. Undoubtedly the greatest concern of people is health. The most severe air pollution occurs in urban areas where large segments of the populace work. Furthermore it is the long time residents of the city—predominately the disadvantaged and the elderly—who usually suffer the greatest exposure to air pollutants. Also, as the levels of education and affluence rise, people appear less and less likely to accept an environment that is asthetically disagreeable, irrespective of whether or not it is harmful. Smoke, haze, unnatural coloration or unpleasant odors are examples of manifestations of atmospheric trace species unacceptable to many persons. Whatever the motivation of individuals the net result appears to be a willingness on the part of people to devote greater attention and financial resources to control of atmospheric pollution. No doubt this will lead to further advances in our ability to measure and understand the nature of combustion processes, the trace species they emit, and the impact of these species on ourselves and our environment.

Historically, interest in combustion processes was largely limited to academicians who were studying combustion and to those engineers involved in the design and development of combustors. The broad nature and the ramifications of the areas discussed above have greatly expanded the need for various degrees of understanding by a much broader spectrum of persons whose interests transcend the subjects of research and design.

As an example of the diversity of involvement it is worth noting that the most significant factors that have influenced the evolution of combustion processes in recent years are not technological but political in nature. Legislation has introduced the emission of trace species as an important constraint on the design and operation of combustion devices. The economic and social dislocations of this legislation are many and far reaching. Among them is the shifting demand for inherently clean burning fuels at the expense of more traditional ones. The development of substitute fuel technologies, such as coal gasification, extraction of oil from shale deposits and the desulfurization of coal and oil, has been accelerated. Likewise additional considerations are being given to alternate noncombustion energy sources such as nuclear, geothermal and solar power. If nothing else the far reaching impacts of legislation regulating combustion processes attests to the basic dependence of our society on such processes.

The legal community has been called upon to clarify and to interpret the intent of this legislation. The great majority of legislators, judges and lawyers are neophytes when it comes to appreciating the details of combustion. They must depend upon information supplied to them by the technical, scientific and medical communities.

Many other groups are involved. Control agency personnel are intimately involved because they often author the detailed rules and regulations governing source emissions. They then enforce them and in doing so may be called upon to recommend control techniques or strategies. The responsibility for funding and directing research and development programs aimed at reducing source emissions may also be theirs.

Transportation and urban planners need to introduce into their plans constraints that will minimize atmospheric pollution. Industrialists and investors need understanding in order to predict the impact of regulations and limitations on industry, on expansion and establishment of new facilities, on profitability and on the market for new products.

The list could go on and on. The point is that combustion is no longer a subject only for technical and scientific journals. Recognizing this it is difficult to draw the line on what subjects to address or not to address in a treatise on the emission of trace substances from combustion processes. Indeed it would not be difficult to ramble *ad infinitum* on all sorts of peripheral issues, many of which have already been identified in this discussion.

1.2 SCOPE OF THIS BOOK

The term *combustion* is used to describe a much broader class of phenomena than will be considered in this book. Combustion or burning basically refers to the rapid oxidation of some substance. Oxidation is an exothermic chemical reaction. Exothermic means heat is evolved as opposed to endothermic which means heat is absorbed. The stipulation that the oxidation be rapid is important. Essentially, when oxidation is rapid, the temperature of the material undergoing oxidation rises rapidly due to the inability of transport processes (conduction, convection, and radiation) to transfer heat to the surroundings as rapidly as it is produced by the oxidation reaction. Thus, a distinction is made between combustion and other types of oxidation that may occur at both a much slower rate and much lower (near ambient) temperatures. The products of these two very different types of oxidation phenomena may or may not be identical.

Many, many substances on this planet burn, and most organic substances fall into this category. The same is true of other substances not ordinarily thought of as combustible. For example, with a few exceptions such as magnesium, ordinarily most metals are not combustible. If, however, they are suspended in air as fine particles, many metals are capable of oxidizing so rapidly that their surfaces become incandescent. Under these circumstances, one would say they burn.

The focus of this book is on relatively few combustion systems and relatively few types of fuel. For example, choosing from those listed in Table I, such a group would include the mobile sources, particularly automobiles, trucks and aircraft, and also stationary sources such as electric power and incineration. These few types of processes and the substances that fuel them are the source of the bulk of all combustion-related air pollutants. There are many, many other combustion processes resulting in the emission of trace species. Among these are explosions, spontaneous combustion of stored materials, welding, smoking, forest fires and sterno fuel. While all of these are of concern in specialized contexts, their overall contribution to air pollution is not of the same order of magnitude as those mentioned earlier. Consequently, the majority of the discussion in the text is oriented toward the first mentioned group.

Another combustion-related subject that will not be discussed is flame retardants. This includes fire extinguishing materials as well as consumer products that are specially treated to retard combustion, such as some types of children's clothing and automobile interiors. To some extent, the principles of a discussion of preventing or retarding combustion are an extrapolation of those that will be presented. Again, however, the emission of trace species from the combustion or perhaps, more appropriately, the smoldering of these substances is not usually considered to present an air pollution problem. Hence, their elaboration is left to other specialized books.

1.3 DEFINITIONS

Fundamentally, there are three distinctly different components in a combustion system: the fuel, the oxidant and the diluents. A fuel containing energy-rich bonds such as the carbon-carbon, carbon-hydrogen and hydrogen-hydrogen bonds is a common source of chemical potential energy. During combustion, such fuels are oxidized and their chemical potential energy is transformed into thermal and sometimes partially into mechanical energy. A list of commonly encountered fuels is presented in Table II. They are subdivided according to their physical and chemical properties. The physical property selected is the phase in which the fuel is initially present at STP [standard temperature and pressure—often chosen as 20°C (68°F) and 760 mm Hg (1 atm) respectively]. These are chosen because the phase (gas, liquid or solid) has an important bearing on many processes and phenomena encountered, among which are the nature of fuel preparation processes and the basic nature of the combustion process itself.

Table II

CLASSIFICATION OF COMMON FUELS

Physical Phase @ STP	Components	Chemical	
		Hydrocarbon (Principal Component)	Other
Gas	Single (pure)	Methane (CH$_4$) natural gas	
		Propane (C$_3$H$_8$) LPG	Producer gas (CO + N$_2$)
	Multicomponent		Coal gas (H$_2$ + CH$_4$)
			Water gas (CO + H$_2$)
Liquid	Single (pure)		Methanol (CH$_3$OH)
	Multicomponent	Crude oil, petroleum distillate products gasoline kerosene fuel oil residual	
Solid	Single		Coke (principally carbon) (derived from coal)
	Multicomponent	Coal Anthracite Bituminous Lignite	Wood, trash, refuse and miscellaneous solid wastes

From a chemical viewpoint, the fuels are subdivided into those that are single component or essentially pure substances and those that are a mixture of several fuel species (multicomponent). Again, this is important primarily because multicomponent fuels in practice often burn quite differently from single component ones. This is particularly true in the combustion of liquid and solid fuels, when the different species in a multicomponent fuel have different vapor pressures. As a result, they may be consumed at different points in both time and space within a flame. This is mentioned here to explain the rationale for the distinction made in Table II, and it will be discussed in much greater detail in Chapter 4.

The distinction between fuels in which hydrocarbons are the principal components and other fuels is included to illustrate the importance of combustion of hydrocarbons or fossil fuels (natural gas, petroleum and coal) to our society. It is no accident that in discussing the principles of combustion the examples chosen and the chemical equations used are largely oriented toward the combustion of hydrocarbon fuels. Once these principles are understood, their extrapolation to other classes of fuel should present no insurmountable difficulties to the reader. For the purposes of an introductory book such as this, the limitation results in a more coherent discussion.

Finally, it should be noted that the classification scheme presented in Table II does not convey any information about the trace or minor components (contaminants) that may be present in a fuel. Natural gas and propane are relatively free of contaminants. Crude oil and the various petroleum distillates derived thereof may contain significant quantities of sulfur and metallic species. The term "significant quantities" in the context of this discussion means they are present at high enough concentrations to represent a potential air pollution problem when the fuel is burned. Coal often contains substantial amounts of sulfur, ash and trace metals.

The second distinct component of a combustion system is the oxidant. The oxidant reacts with the fuel during combustion to transform the chemical potential energy stored in the fuel into thermal energy. Most commonly, the oxidant is molecular oxygen, a constituent of air. However, there is no reason why the oxidant is limited to a gaseous substance like oxygen. It may be a liquid or a solid, as is the case with explosives and propellants. However, since oxygen is the most commonly used oxidant, the discussion throughout this book will emphasize the combustion of fuels in air.

Since the focus of this book is on a subject related to air pollution, it is worth mentioning the distinction between the use of the term *oxidant* here and its use in discussions of air pollution. Here the oxidant is the chemical species that combines with the fuel to release the stored energy. In discussions of air pollution, oxidant usually denotes species with a much greater propensity than oxygen to oxidize other species. Ozone, an example of such a substance, exists at concentrations of 1 part per million (ppm) or less whereas the atmospheric concentration of oxygen is approximately 210,000 ppm. The need to express the concentration of trace species has led to the use of units such as ppm. The reader unfamiliar with this and other conventions used for expressing trace concentrations should refer to the discussion in Appendix A.

The third and final component that may be present in a combustion system is the diluent. A diluent is a substance that does not participate chemically in the combustion reaction either as a fuel substance or as an oxidant. It is physically present and often does influence the combustion process. For example, diluents have heat capacity and while they do not make a positive contribution to the total energy released, they do act as a thermal sink and limit the temperature rise achieved by combustion. A diluent can be thought of as a substance that participates principally in the physical aspects of the combustion process.

Nitrogen, which comprises greater than 78% of air, is the most common diluent. Components of air such as water vapor and the inert gases are also diluents, as are inorganic ash compounds present in coal and petroleum products. The term *inert* should not be taken too literally. For example, at high temperatures a small portion of the so-called inert nitrogen does enter into the overall chemical reactions to produce one or more oxides of nitrogen, which are among the most important emission products of combustion. Other inert substances may also undergo chemical reactions. Water vapor is a product of complete combustion; it is neither a fuel nor an oxidant. However, when it is present in the air supplied to the combustion process, it may alter the process physically by virtue of its behavior as a thermal sink and chemically via alteration of the high temperature equilibrium distribution of the reacting gas mixture.

Molecular oxygen was previously identified as an oxidant. If, however, the quantity of oxygen present is in excess of that needed to completely oxidize the fuel, then the excess oxygen will act as a diluent. That is, its participation in the combustion process will be physical but not chemical. The same may be said for any unreacted excesses.

In most applications, inerts are present because it is more convenient to use air as an oxidant than to separate the oxygen from the nitrogen. There are, however, special situations such as the burning of some solid explosives or propellants where inert substances are purposely added to limit the temperatures attained or the rate of burning. In these cases, the inert substance may be referred to as a coolant.

The various types of combustion that will be discussed in this book appear in Table III. There are premixed and diffusion flames, monopropellant and propellant combustion, and explosions. Their occurrence is classified according to the initial states of the fuel and the oxidant. This scheme of classification is useful since the initial states of the fuel and oxidant are observations that are readily made when examining a practical combustion system. With the exception of the premixed flame for Case I, which is discussed in detail in Chapter 2, all of the other cases are discussed in Chapter 4.

Table III
TYPES OF COMBUSTION

Case	Initial State		Fuel and Oxidant	
	Fuel	Oxidant	Mixed	Separate
I	Gas	Gas	Premixed flame	Diffusion flame
II	Liquid	Gas	Premixed flame	Diffusion flame
III	Liquid	Liquid	Monopropellant combustion	— — —
IV	Solid	Gas	— — —	Diffusion flame
V	Solid	Solid	Propellant combustion, explosion	— — —

The distinction between the fuel and oxidant being mixed or not, which is made in the two right-hand columns of Table III, refers to their dispersion on a molecular scale rather than on a macroscopic or even a microscopic one. The distinction is readily illustrated by Case II. Initially the fuel is a liquid and the oxidant a gas; for example, the burning of fuel oil in air. Normally, the oil is atomized to facilitate combustion. If the droplets are allowed to vaporize and the vapors to mix with the surrounding air before combustion is initiated, burning takes place in a premixed flame. But, when flames envelop each droplet and consume the fuel vapors before they mix with the air, then burning is said to take place in a diffusion flame, that is, a flame in which countercurrent diffusion is the source of the fuel and oxidant reactants for the combustion reaction.

The subjects of propellant, monopropellant and explosive reactions are discussed briefly in Chapter 4. A propellant is a solid or liquid fuel used to propel a rocket, guided missile, etc. The term monopropellant is usually limited to a liquid mixture of fuel and oxidizer. An explosive is any substance or mixture of substances that, on impact or by ignition, reacts by a violent expansion of gases and thereby results in the liberation of relatively large amounts of thermal energy. Technically, of course, any premixed combustible system can be an explosive. The term as used here is restricted to solids and mixtures thereof; for example, nitrocellulose and black gun powder. In all these cases, the reactants are usually premixed to facilitate a rapid reaction rate. The lack of entries in some spots does not imply the impossibility of such a combustion system, but only that such a case is not addressed in this book.

At this point, the different kinds of reactions or processes encountered in this book need to be mentioned. Oxidation has already been introduced. There are others, for example, pyrolytic reactions and catalytic and noncatalytic reactions. Distinctions also need to be made between homogeneous and heterogeneous and between flame and nonflame processes.

Oxidation, in the limited sense that is used throughout most of this book, means the combination of a fuel substance with molecular oxygen. Distinctions are made between high and low temperature oxidation, *i.e.*, oxidation occurring in a flame *vs.* oxidation occurring at a substantially lower temperature either before or after the flame. The rationale behind this is basically one of distinguishing between two processes that occur by quite different mechanisms.

The term *pyrolysis* denotes a class of reactions that occur in the absence of oxygen. Pyrolysis, then, refers to reactions of the fuel rather than reactions of the other two components (oxidant and inerts) of a combustion system. Basically, fuel substances undergo chemical reactions of one sort or another when the temperature is raised sufficiently. If there is oxygen present, they oxidize. If no oxygen is present, reactions occur between different fuel molecules or fragments thereof, and this is pyrolysis. It is an important class of reactions and can occur in both premixed and diffusion flame systems. Pyrolytic reactions often significantly influence both the kind and concentration of trace species produced by combustion.

A distinction is made between homogeneous and heterogeneous reactions. A reaction is homogeneous if it takes place in a single phase. It is heterogeneous if it requires the presence of at least two phases to proceed at the rate it does. It is immaterial whether the reaction takes place in one, two or more of the phases, or at an interface between them. All that is important is that at least two phases are necessary for the reaction to proceed.

The distinction between flame and nonflame processes is a complex one that is considered in detail in Chapter 2. One's first inclination is to think of flames *vs.* nonflames in terms of reactions that are luminous *vs.* those which are not, respectively. This is not entirely adequate since there are many physical and chemical reactions that are luminous but are not flames, for example, fluorescence and phosphorescence. To complicate matters further, there are flames that under ordinary conditions do not appear luminous to the human eye. However, at this point, it would be confusing to become involved in the chemical distinctions of flame and nonflame processes. It will suffice to note that flames are zones in which rapid oxidation reactions occur, which in turn result in large discontinuities in composition and temperature. The flame zone has rather sharp spatial boundaries and the relatively high temperatures produced within the zone often result in the emission of visible radiation. Strictly speaking, since the term *phase* implies uniform temperature, pressure and composition throughout, a flame is not a homogeneous process.

For purposes of this book, the term heterogeneous is used only to describe the noncatalytic and catalytic reactions occurring at gas-solid interfaces. A noncatalytic reaction of this type, the oxidation of particulate matter in the hot gases behind the flame, is introduced in Section 2.8. A number of important catalytic reactions are discussed in Section 3.4. In the cases discussed, the solid surface is the catalyst and is used to promote the rate of desired reactions in the effluent from combustion processes without undergoing any net change itself.

Finally, the reaction products of combustion need to be considered. The products of oxidation are referred to as partial and complete oxidation products. The use of the terms *partial* and *complete* makes reference to the oxidation state of the products. Examples of completely oxidized species are carbon dioxide and water; examples of incompletely oxidized species are carbon monoxide and formaldehyde. It is possible for these species to continue to react with oxygen if the correct conditions prevail, that is, if the temperature and oxygen concentration are high enough for the reaction to proceed in the time available.

High molecular weight hydrocarbon and particulate species may be produced by pyrolytic reactions. Trace species in the fuel may be oxidized, resulting in the formation of species such as sulfur dioxide, sulfur trioxide, and various metallic oxides. Finally, it should be mentioned that at the high temperatures achieved in flames, reactions occur between the oxygen and nitrogen in the air and produce nitric oxide.

The above-mentioned species are by no means an exhaustive list of the trace species emitted by combustion processes but rather some of the more commonly encountered ones. The reader who is not familiar with these chemical species will probably find the discussion in Appendix B of value.

1.4 QUANTIFICATION OF COMBUSTION

The most useful type of technical analysis of any physical or chemical phenomenon is a quantitative one. Generally, measurements and other observations of a phenomenon are made, and then based on these, a quantitative description is developed. If the process is well understood, a reasonably accurate theoretical model can usually be developed. If this is not the case, or if the process is extremely complex, an empirical correlation of important variables may be more useful.

It is assumed that the common physical properties of a system such as temperature and pressure are familiar to the reader. There are, however, some specialized terms and means of characterizing combustion systems that may be unfamiliar to the reader, and these will be discussed in this section.

A stoichiometric mixture of fuel and oxidant is one in which the quantity of oxidant present is just sufficient to completely oxidize the fuel. Therefore a stoichiometric mixture is one in which the ratio of fuel to oxidant is theoretically correct for complete oxidation. For example, the simplest hydrocarbon is methane. Stoichiometric combustion requires two moles of oxygen for each mole of methane.

$$\underset{\text{Fuel}}{CH_4} + \underset{\text{Oxidant}}{2O_2} \rightarrow \underset{\text{Combustion Products}}{CO_2 + 2H_2O} \tag{1}$$

Since the oxidant is usually present as a component of air, Equation (1) may alternatively be written as follows:

$$\underset{\text{Fuel}}{CH_4} + \underset{\underset{\text{Air}}{\underbrace{\text{Oxidant + Diluent}}}}{2\left[O_2 + \left(\frac{79}{21}\right) N_2\right]} \rightarrow \underset{\substack{\text{Combustion} \\ \text{Products}}}{CO_2 + 2H_2O} + \underset{\text{Diluent}}{2\left(\frac{79}{21}\right) N_2} \tag{2}$$

In Equation (2) it is assumed that the air is composed of 21% (by volume) oxygen with the balance, 79%, being nitrogen. This is only an approximation, but it is sufficient for many combustion computations.

If the quantity of oxygen in the combustible mixture is in excess of that required for complete combustion of the fuel, then a portion of the oxygen does not react and appears in the exhaust. If $(2 + a)$ moles of air rather than 2 moles are present then Equation (2) becomes:

$$\underset{\text{Fuel}}{CH_4} + \underset{\text{Air}}{(2 + a)\left[O_2 + \left(\frac{79}{21}\right) N_2\right]} \rightarrow \underset{\substack{\text{Combustion} \\ \text{Products}}}{CO_2 + 2H_2O} + \underset{\text{Diluents}}{(2 + a)\left(\frac{79}{21}\right) N_2 + aO_2}$$

$$\tag{3}$$

A combustible mixture that contains excess oxidant is referred to as "lean." The converse case where there is insufficient oxidant for complete oxidation of the available fuel is referred to as "rich." Rich combustion results in the production of a mixture of partial and complete oxidation products. Carbon dioxide and water are complete oxidation products for the oxidation of methane. Carbon monoxide and hydrogen are products of partial oxidation, or alternatively stated, products of the incomplete oxidation of methane.

When the combustible mixture contains excess fuel, the relative proportions of CO_2, CO, H_2O and H_2 that are produced depend upon the degree of richness and the temperature and pressure conditions to which the working fluid is subjected as it undergoes combustion. Thus it is not possible to write a balanced chemical equation analogous to (3) for rich combustion without more complete knowledge of these additional factors.

Thus far the discussion has been limited to combustion of methane. A generalized formula for combustion of hydrocarbons in air is:

$$CH_y + nO_2 + n\left(\frac{79}{21}\right)N_2 \rightarrow a\,CO_2 + (1-a)CO + b\,H_2O + \left(\frac{y}{2}-b\right)H_2 +$$

$$\left[n - a - \frac{1-a}{2} - \frac{b}{2}\right]O_2 + n\left(\frac{79}{21}\right)N_2 \tag{4}$$

In this equation the hydrocarbon composition has been normalized with respect to the number of carbon atoms. For methane with only a single carbon atom $CH_y = CH_4$. For propane (C_3H_8), $CH_y = CH_{8/3}$ or $CH_{2.67}$ and in the case of benzene (C_6H_6), $CH_y = CH_{1.0}$. Acetylene (C_2H_2) has the same value of y as benzene.

It is necessary to be able to express not only that a combustible mixture is rich, lean or stoichiometric, but also how rich or lean it is. Fuel-air or air-fuel ratios and equivalence ratios are the most common ways of quantitatively expressing the richness or leanness of a mixture of fuel and oxidant. If n_f is the moles of fuel, M_f the molecular weight of fuel, and n the number of moles of oxygen per mole of fuel as shown in Equation (4), then the fuel-air ratio of the mixture is either

$$\left(\frac{F}{A}\right)_{mol} = \frac{n_f}{n\left[1 + \left(\frac{79}{21}\right)\right]} \tag{5}$$

or

$$\left(\frac{F}{A}\right)_{wt} = \frac{n_f M_f}{n\left[1 + \left(\frac{79}{21}\right)\right]29} \tag{6}$$

where the molecular weight of air is taken as 29.

Obviously air-fuel ratios are reciprocals of fuel-air ratios, $(A/F)_{mol} = (F/A)_{mol}^{-1}$ and $(A/F)_{wt} = (F/A)_{wt}^{-1}$. The numerical value of (A/F) or (F/A) ratios at the stoichiometric (or theoretical) point depends upon the fuel composition, that is, upon the value of y in CH_y.

Often it is desirable to compare the richness or leanness of combustion for different fuels. The equivalence ratio, $\emptyset$, is convenient for this type of comparison, and it may be defined as the quotient of the actual fuel-air ratio divided by the stoichiometric fuel-air ratio.

$$\emptyset = \frac{(F/A)\ actual}{(F/A)\ stoichiometric} \tag{7}$$
$$(theoretical)$$

or, alternatively, in terms of air-fuel ratios

$$\emptyset = \frac{(A/F)\ actual}{(A/F)\ theoretical} \tag{8}$$

Again, for a stoichiometric mixture $\emptyset = 1.0$. However, it should be recognized that even though the equivalence ratio defined in Equations (7) and (8) have identical values at stoichiometric, they are not identical for "off-stoichiometric" (rich or lean) mixtures because (F/A) and (A/F) ratios are reciprocals. This is illustrated in Table IV.

Table IV

EQUIVALENCE RATIO

Definition of $\emptyset$	Mixture	Value
Per equation (7)	Rich	$\emptyset > 1$
	Stoichiometric	$\emptyset = 1$
	Lean	$\emptyset < 1$
Per equation (8)	Rich	$\emptyset < 1$
	Stoichiometric	$\emptyset = 1$
	Lean	$\emptyset > 1$

Care must be exercised in the use of (A/F), (F/A) and equivalence ratios. First, either mol or weight ratios may be employed to calculate (A/F) or (F/A) ratios. Second, equivalence ratios may be calculated from either (F/A) or (A/F) ratios and with the exception of the stoichiometric point the results are not interchangeable. The convention chosen should always be clearly stated by the writer or ascertained by the user and be maintained throughout the entire calculation.

The concentrations as volume per cent of the principal combustion products of hydrocarbon fuels for equivalence ratios from 0.6 to 1.3 are shown in Figure 1. The equivalence ratio was computed from air-fuel

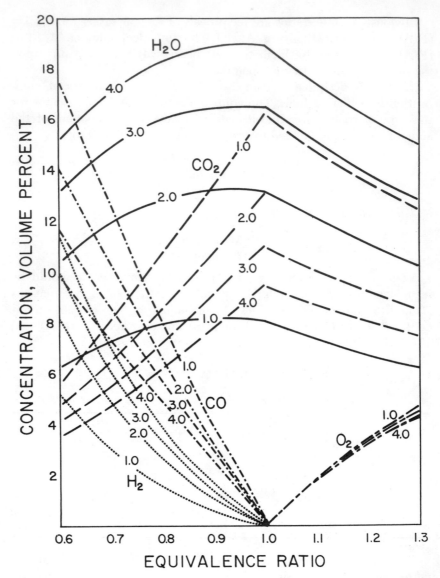

Figure 1. Concentration as volume per cent of the principal combustion products of hydrocarbon fuels. Equivalence ratio computed from air-fuel ratios. Parametric values, y, are the hydrogen-carbon ratios in the fuel. Adapted from information in Reference 1.

weight ratios. Therefore, $\emptyset < 1$ corresponds to rich combustible mixtures. Note that the relative concentrations of CO_2, CO, H_2O and H_2 depend upon both the value of $\emptyset$ and fuel composition as was men-

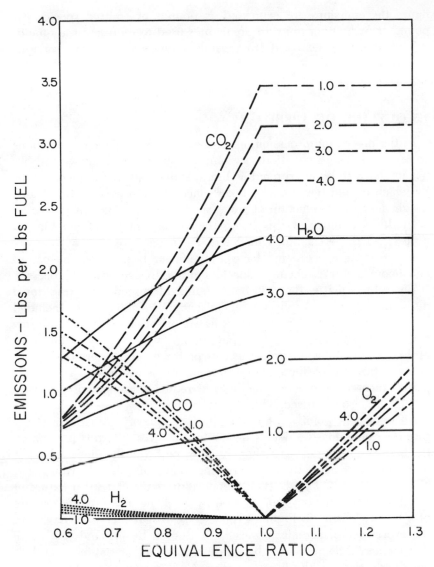

Figure 2. Weight of emission products as pounds per pound of fuel for hydrocarbon combustion. Equivalence ratio computed from air-fuel ratios. Adapted from information in Reference 1.

tioned earlier. This figure is based upon calculations made by D'Alleva[1] who utilized the water-gas equilibrium relationship to establish the concentrations of combustion products on the rich side of stoichiometric. The same information is shown in Figure 2. However, in this figure the emissions are expressed as pounds of emitted species per pound of fuel. Other units are sometimes used to represent the quantities of substances emitted from combustion processes, for example, pounds of a species per million BTU of energy released or pounds of a species per ton of fuel consumed.

1.5 OUTLINE OF APPROACH

In the preceding discussion, a number of overall chemical reactions were stated. These equations are only accounting statements and are not intended to be interpreted as literal statements of the mechanisms by which combustion actually takes place. They provide little, if any, insight into the production of trace species by combustion processes.

In the remainder of this book, the various facets of combustion processes will be examined in detail. Particular attention will be devoted to those aspects that have a significant bearing on the nature and amounts of trace chemical species emitted by combustion processes.

The subject matter that will be introduced transcends the traditional disciplinary limits of physics, chemistry and engineering. Included are such diverse physical processes as vaporization, mixing, fluid flow and conduction, convection and radiation. A wide variety of chemical processes are discussed from the viewpoints of both thermodynamics and kinetics. In addition to the aforementioned subjects, which are largely theoretical, practical methods of reducing the emission of trace species are also discussed. This is engineering.

An important point must be emphasized. Theory provides a structure for examining the details of combustion. Usually, this is done on the basis of an average case. For the most part, the theories evolved to date give insufficient attention to the fact that reality is a distribution of all possible states. That is, they do not address the statistical nature of the process. An analogy can be seen by referring ahead to the picture of a Bunsen burner in Figure 4 (page 23). Most readers will have had the experience of actually observing a Bunsen burner and will recall that the actual flame flickers both in intensity and spatial location. It is not as invariant as the figure implies. Now it is possible to describe theoretically a process in terms of the variables that have an important effect on the emission of trace species. It is also possible from an engineering point of view to design and operate a combustion process in a way that minimizes its emission by selecting an optimum set of values for the variables. In fact, precise control of such variables is one of the most fundamental approaches to reducing the emission of trace species. A point is reached, however, where more precise control is not possible because the variations are statistical in nature and beyond the intervention of humans. There is a certain minimum level of pollutants that corresponds to distribution of nonoptimum states that are actually occurring due to statistical variations. Reducing emissions below this level is not possible without alteration or modification of the actual process. This is mentioned because the theories that exist are incomplete and can easily lead the reader to believe the impossible can be accomplished.

The order of discussion is as follows. First, premixed combustion will be considered in some detail. This approach is chosen because in a premixed system the overall combustion rate is determined by chemical factors alone. The general principles developed in the premixed flame discussion are applicable to the more complex processes occurring in diffusion flames. In the latter case, the situation is complicated by the occurrence of additional processes, such as the mixing of the fuel and oxidant, that may be rate controlling. Subsequent treatment of these more complex combustion systems is then undertaken without the necessity of repetitious discussion of basic principles. It is hoped that this approach will be beneficial by demonstrating both the similarities existing between seemingly very different processes and the differences existing between similar ones.

Only minimal consideration is given to the details of specific applications in the chapters that follow. The reasons for this are threefold. First, different terminology has evolved in different areas of application. For example, quite different terms are used to describe the same phenomena in the automotive and central power generating fields. The author has tried to adopt a terminology that is not redundant and that is readily translatable into different areas of application. Second, given the present situation where technology is rapidly evolving, a discussion of applications is likely to be dated rapidly. Third and finally, the applications of interest are so diverse, as is apparent from Table I, that an individual working in a particular field is more likely to be intimate with the details than is the author. For the reader who is not expert in a particular area, the author has tried to identify representative literature on different applications. This is presented in Chapter 5.

REFERENCES

1. D'Alleva, B. A. "Procedure and Charts for Estimating Exhaust Gas Quantities and Composition," General Motors Research Publication No. 372 (May 15, 1960).

PREMIXED FLAMES

2.1 INTRODUCTION

A method of classifying combustion according to whether or not the fuel and the oxidant are mixed was set forth in Table III. In most all combustion systems, of course, the fuel and the oxidant are *initially* unmixed. What is important, however, is whether on a molecular scale they are mixed or are not mixed *prior to entering* the flame, *i.e.*, the combustion zone. The rationale for making this distinction is based on kinetic considerations. When a system is mixed on a molecular scale, the overall combustion rate is determined by chemical factors, whereas when the fuel and the oxidant are not so mixed, physical processes such as mixing are often rate controlling.

A familiar example of premixed combustion is the burning of gasoline in an ordinary automobile engine. The fuel, which is initially liquid, is mixed with air in the carburetor. Vaporization is ordinarily completed in the intake manifold or in the cylinder before a spark initiates combustion. Examples such as this, which are probably familiar to most readers, are introduced from time to time to assist in visualization of the kinds of processes being discussed. For a detailed treatment of combustion processes in these practical applications, the reader is referred to Chapter 5.

Other examples of premixed combustion systems are the laboratory Bunsen burner, which will be considered in more detail shortly, and the explosion of a fuel vapor-air mixture. If a pool of flammable liquid evaporates and its vapors are allowed to mix with air, a combustible vapor-air mixture may result. If a source of ignition is present (a spark, a hot light bulb, etc.) the mixture may be ignited resulting in a flash fire or explosion.

This latter example illustrates why the use of premixed combustion is largely limited to relatively small devices (*i.e.*, automobile engines and laboratory burners). It is not particularly desirable to confine large quantities of combustible gases. There is the possibility of a "flash-back" from the flame to the fuel-air supply resulting in a destructive explosion. Diffusion flames, which will be discussed in Chapter 4, avoid this possibility by maintaining the fuel and the oxidant in an unmixed state prior to their entry into the combustion zone.

The discussion in this chapter is limited to the combustion of pre-mixed gaseous fuel-oxidant mixtures. This does not mean that the fuel is gaseous at STP. It may be a liquid as in the case of the automobile engine. The only implication is that the fuel-air mixture is gaseous when it approaches the combustion zone. Premixed liquid and solid fuel-oxidant systems (propellants) will not be considered here. These are rather specialized situations and will be briefly considered at the end of Chapter 4.

The convention adopted of limiting the use of the term *premixed* to systems that are intimately mixed on a molecular scale also excludes the consideration at this point of finely divided liquid and solid suspensions in air. On a macroscopic or even a microscopic scale, these systems are mixed. However, from a mechanistic point of view, as long as the fuel is a condensed phase, the process of combustion will usually occur as a diffusion flame enveloping each particle or droplet. Discussion of this phenomenon is reserved for Chapter 4.

2.2 THE PREMIXED FLAME

Figure 3 depicts the major processes that occur in a premixed flame. Air and gaseous fuel enter from the left. They are intimately mixed

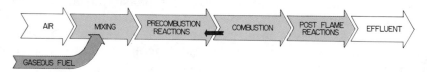

Figure 3. Major processes occurring in a premixed flame.

to produce a fuel-air mixture in which the two are uniformly distributed on a molecular scale. Premixing does not guarantee homogeneity. The relative proportions of a mixture, that is its richness or leanness, may vary from time to time at a given point or from point to point at a given time in a premixed fuel and air mixture. This mixture undergoes a series of chemical reactions. First are the precombustion reactions that occur prior to combustion itself. Second is the combustion process when the rapid conversion of chemical to thermal energy occurs. Finally, there are the post flame reactions, or as the name implies, those reactions that take place subsequent to the flame in the burned gases. Finally, the combustion products which are referred to as either the effluent or the exhaust are discharged into the atmosphere. Diagrammatic representations such as this are introduced throughout this chapter and will illustrate the sequence of important processes in time and space.

An example of a system that approximates premixed combustion is the simple Bunsen burner, shown in Figure 4. The fuel and air enter near the base. Mixing occurs as the gases flow upward through the burner stem and produce a fuel-air mixture that is more or less premixed prior to entering the flame zone. Combustion occurs in the flame zone and the products of combustion mix with the surrounding air.

Returning for the moment to Figure 3, as the fuel-air mixture approaches the high temperature combustion zone, two important trans-

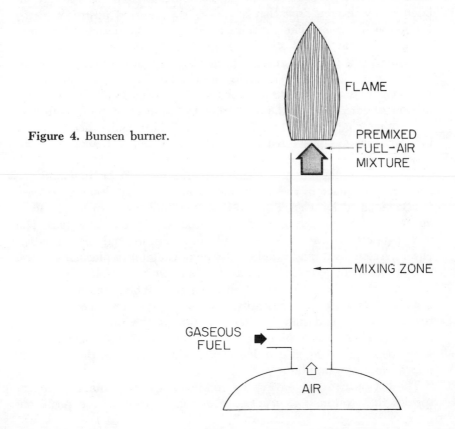

Figure 4. Bunsen burner.

FLAME

PREMIXED FUEL-AIR MIXTURE

MIXING ZONE

GASEOUS FUEL

AIR

port processes occur.[1] Both heat and chemical species are transported upstream from the high temperature combustion zone to the unburned fuel-air mixture. The term upstream refers to flow which is countercurrent to the bulk flow. These countercurrent transport processes are depicted by the small solid arrow in Figure 3.

The countercurrent flow of heat occurs because of convective diffusion or radiant transport. The transport of thermal energy upstream from the combustion zone increases the temperature of the fuel-air mixture. This together with the transport of high energy reactive chemical species in the same direction promotes the occurrence of chemical reactions in the fuel-air mixture. These reactions that occur prior to the combustion zone are termed *precombustion reactions* as mentioned earlier and noted in Figure 3.

The preceding discussion might be summarized with the statement that the precombustion and combustion zones are thermally and diffusionally coupled. The net result of this coupling is that the reactions that occur downstream in the combustion zone are strongly influenced by the reactions that have already occurred upstream in the precombustion zone. The converse is also true. The nature of the precombustion reactions is strongly influenced by the combustion reactions since it is the high energy species that are produced in the combustion zone and diffuse upstream which initiate the precombustion reactions. It is not possible to alter the process in either the upstream or downstream zones without altering the process in the other. It is this strong coupling that distinguishes a flame from other types of homogeneous gas phase reactions that also proceed at high temperatures.

The bulk of the energy released by reaction of the fuel and the oxidant takes place in the combustion zone. This rapid energy release is accompanied by a rapid, almost discontinuous, rise in temperature and a simultaneous change in the composition of the reacting gas. The high temperature causes widespread bond rupture and the formation of ions, free radicals and simple diatomic or triatomic molecular species.

It is common to equate the zone of a flame that is visible or luminous (to the human eye) with the combustion zone. This is not an entirely correct comparison for luminosity may well persist into the post-flame reaction region and under certain conditions the precombustion reactions may also emit visible radiation. Furthermore, the combustion of some fuels such as hydrogen does not produce radiation that is visible under ordinary conditions.

The hot burned gas leaving the combustion zone contains chemical species that continue to react as these gases cool. These postflame reactions play a significant role in determining the chemical constituents ultimately present in the effluent. References 2, 3, 4 and 5 present a more detailed discussion of the general subject of combustion.

In the following sections, each of the major processes shown in Figure 3 will be examined in greater detail.

2.3 PRECOMBUSTION REACTIONS

The term *precombustion* denotes those reactions that occur prior (in time) or upstream (in space) of the ignition and combustion of a fuel-air mixture. In this section the nature of these reactions and the implications that their occurrence has with respect to emissions products from combustion processes are discussed.

A short notation for a hydrocarbon molecule is RH where R denotes a complex group ranging from the methyl group, $CH_3^{\bullet}$, at the simplest to a much more complex group expressed generally as C_xH_y. The H in the RH represents any one of the hydrogen atoms present in the molecule. Direct reaction between a hydrocarbon molecule RH and an oxygen molecule

$$RH + O_2 \rightarrow R^{\bullet} + HO_2^{\bullet} \tag{9}$$

is not a very probable event under ordinary conditions. This elementary reaction is very endothermic (50–60 kcal/mole) and its activation energy is high. The rate of Reaction (9) is very small even at temperatures in the range of 900–1000°F. This is near the upper limit of temperatures occurring in the precombustion zone as a result of countercurrent transport of thermal energy from the combustion zone. The countercurrent mass transport of high energy chemically active species such as hydroxyl radicals ($OH^{\bullet}$) from the combustion zone to the precombustion zone is responsible for the initiation of free radicals in this latter zone. For example, hydroxyl radicals react with hydrocarbons to produce alkyl, allyl and aryl radicals as follows:

Reactions (10)–(12) have low activation energies. For example, the activation energy of Equation (10) is only 2 kcal/mole as compared to

$$RH + {}^{\bullet}OH \rightarrow R^{\bullet} + H_2O \tag{10}$$

alkane   alkyl radical

$$H_2C = CHCH_2R + {}^{\bullet}OH \rightarrow H_2C = CH\overset{\bullet}{C}HR + H_2O \tag{11}$$

olefin   allyl radical

$$C_6H_5 - CH_3 + {}^{\bullet}OH \rightarrow \qquad C_6H_5 - CH_2^{\bullet} + H_2O \tag{12}$$

aromatic   aryl radical

50 to 60 kcal/mole for direct reaction of molecular oxygen with a hydrocarbon. This explains the relatively large reaction rates of Reactions (10)–(12) relative to that for Reaction (9).

In order to better visualize the processes being discussed here, consider the combustion of a premixed gas flowing in a pipe. This is illustrated in Figure 5. At the top is the sequence of processes occurring in

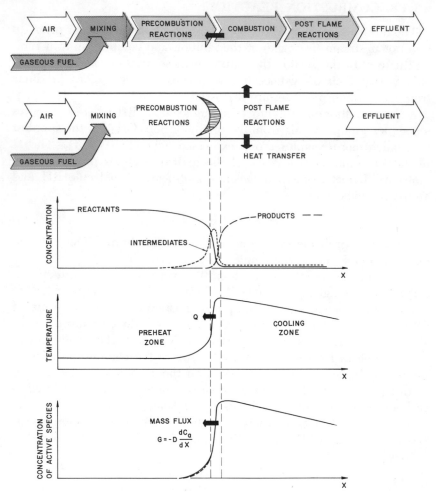

Figure 5. Premixed combustion of a gaseous mixture flowing through a pipe.

a premixed flame, previously introduced in Figure 3. Fuel and air enter from the left and after mixing they undergo precombustion reactions of the type presented in Equations (10) through (12). These reactions are promoted by the transfer of heat and active species from the combustion to the precombustion region as is illustrated by the lower two parts of the figure. Figure 5 is only intended to convey general relationships. No units are shown on the abscissas or the ordinates. Numerical values of temperature and concentration depend upon many factors that are unique to each particular situation. It might be noted, however, that the temperature rise through the combustion zone is very large, and the temperature gradient (dT/dx) may be of the order of $10^5°C/cm$. The magnitude of flux of active species depends upon both the concentration gradient of the active species within the combustion zone, which like the temperature gradient is large, and upon the diffusion coefficient, D, which increases with temperature. Thus, a large concentration gradient combined with high temperatures in the zone where the active species originate produces a not insignificant concentration of active species in the precombustion zone. This is illustrated by the concentration profile for an active species in the lower part of Figure 5.

Returning to the discussion of precombustion reactions, the alkyl radical R• produced in Reaction (10) can react directly with molecular oxygen to form a peroxy radical.

$$R^• + O_2 \rightarrow ROO^• \tag{13}$$

The activation energy for this reaction is typically 2–6 kcal/mole depending upon the molecular structure of the R• group. In turn, the peroxy radical formed in Reaction (13) can undergo a variety of subsequent reactions. For example, it may abstract a hydrogen atom from another fuel molecule to produce a peroxide and another alkyl radical.

$$ROO^• + RH \rightarrow ROOH + R^• \tag{14}$$

Again, the activation energies for reactions of this type are not large. Alternatively, the peroxy radical may undergo an internal isomerization wherein one of the hydrogens from the alkyl part of the peroxy radical is transferred to complete the peroxide group.

$$ROO^• \rightarrow {}^•ROOH \tag{15}$$

When alternative reaction paths such as those shown in Equations (14) and (15) exist for reaction of a species like ROO•, a portion of the ROO• radicals may react via one path and the remainder by other paths. Also, the group denoted as R may be a complex structure containing many, perhaps a score or more, of hydrogen atoms. Because the activation energy for Reaction (8) is very low, the probability for abstraction of any given hydrogen atom is about equal to that for the abstraction of any other hydrogen atom. Thus a variety of different alkyl radicals, $R^\bullet$, which differ only in the position of the unpaired electron, are produced. Subsequent addition of oxygen to these different radicals will produce different peroxy radicals. In some cases the structure of the radical is such that the terminal oxygen of the peroxy radical can readily form a bridge with a remote hydrogen of the alkyl radical. In this case isomerization can occur. If the structure of the peroxy radical is such that bond distortion is required for a bridge to form, then the activation energy for that particular reaction is large and peroxide formation via hydrogen abstraction from another fuel molecule is more probable.

The multiplicity of reaction paths that exists due both to the complexity of the reactant molecules RH and the low activation energy for many of the reactions cited helps to explain the large number of different reaction products and intermediates produced by precombustion type reactions.

The organic peroxides formed via Reactions (14) and (15) contain a weak oxygen-oxygen bond. As the reaction mixture approaches the flame, bond rupture will occur.

$$ROOH \rightarrow RO^\bullet + {}^\bullet OH \tag{16}$$

This is a chain branching reaction. Chain branching, of course, is conducive to rapid increases in overall reaction rate, and ultimately culminates in ignition. Alternatively, prior to ignition the alkoxy radical $RO^\bullet$ may react to produce any number of other species including aldehydes ROH, ketones ROR, alcohols ROH and O-heterocyclics.

The preceding discussion provides, at best, a qualitative appreciation of the different types of reactions and reaction products that may occur in the precombustion region. Numerous investigations have been carried out to study the mechanism and products of precombustion type reactions. These investigators call the reactions low-temperature oxidation, slow oxidation and cool flames. The details of these studies are beyond the scope of the present discussion. Should the reader desire to study the reaction mechanisms and products of different fuels and mixtures thereof, information can be found in References 6, 7, 8.

Significance of Precombustion Reactions

There are two reasons for examining the nature of precombustion reactions in detail. First, precombustion reactions influence the combustion process and second, they may have a marked influence upon the nature of hydrocarbon and particulate emission from combustion processes.

Precombustion reactions modify the fuel-air mixture. The reactant mixture entering the flame is not identical to that produced by mixing the fuel and the air. It is one that has been modified chemically by the occurrence of precombustion reactions prior to ignition. Ignition occurs when the temperature of the mixture increases to the point where chain branching reactions produce a rapid release of energy and a corresponding rapid change in mixture composition. The temperature increase that precedes ignition may be produced in a number of ways. It may result from compression of the gas mixture, by contact of this mixture with a hot object such as a spark, a hot spot or a pilot flame, or by heat transfer from the flame itself as discussed earlier. Return for a moment to the center part of Figure 5. The reactants are fuel and oxygen. The intermediates include all of the various species mentioned in the previous section $R^\bullet$, $RO^\bullet$, $ROO^\bullet$, $ROOH$, RCH, ROR, ROH etc.

$$RCH \parallel O$$

To the extent that the mixture is ignited, all of these intermediate species are subsequently oxidized in the flame to produce CO_2, H_2O, and in the case of a rich flame, CO and H_2.

Situations in which the reacting mixture does not ignite, or where some portion of it fails to ignite, are of particular interest here because the intermediates species produced by the precombustion reactions will not necessarily be completely oxidized, and they or other species resulting from their subsequent reaction may appear in the effluent. There are a number of reasons why a fuel-air mixture may fail to ignite. To begin with the ratio of fuel and air may be sufficiently removed from the stoichiometric ratio so that it falls outside the limits of flammability. In this case the large excess of either fuel or air prevents the temperature from rising to the ignition point in spite of the precombustion reactions that occur. These reactions produce a chemically modified gas mixture containing partially oxidized organic molecules. The principal concerns here are to what extent the fuel has oxidized and what partial oxidation products are formed.

Of perhaps even greater interest than the case discussed above is the situation in which only a portion of the fuel-air mixture is consumed by the flame. This is of particular interest because it is a common occurrence in many combustion systems. Consider the flame propagating in the tube as shown in Figure 5. Inflammation of the flowing gases occurs so long as the heat released in the flame per unit volume per unit time is sufficient to raise the mixture to its ignition point. Now, to prevent material problems most surfaces are maintained at temperatures well below those encountered in a flame. Consequently, the heat

transfer to the surface prevents the flame from propagating all the way
to the surface. The temperature gradient in the film of gas immediately
adjacent to the wall is illustrated in Figure 6. This boundary layer is

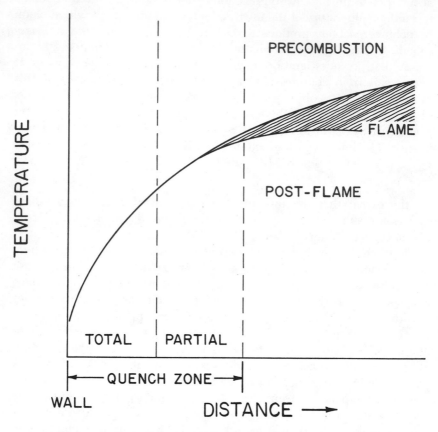

Figure 6. Quench layer in the vicinity of a cooled surface.

sometimes referred to as the quench layer or quench zone implying
that combustion reactions are quenched in this region. Actually the
quenching is only partial. The layer nearest the wall may indeed be
totally quenched if the wall temperature is maintained at a low enough
value. However, due to the transfer of thermal energy and of reactive
species from the flame to the quench layer, precombustion type reactions do occur in the higher temperature portion of the layer.

Incomplete reaction of a combustible mixture due to thermal quenching near the walls is illustrated in Figure 7. Assume the velocity of the

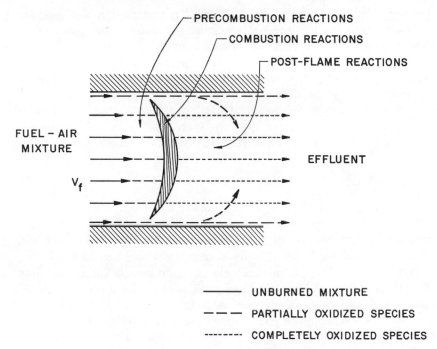

Figure 7. Reaction regimes for a combustible mixture flowing in a pipe with cooled walls.

unreacted fuel-air mixture entering from the left is just sufficient to cause the flame to remain stationary within the tube. That is, the inlet velocity equals the flame velocity V_f. Precombustion reactions partially oxidize the fuel-air mixture as it approaches the stationary flame. The bulk of this partially oxidized mixture is completely oxidized as it passes through the flame zone. However, that portion of the gas in the quench zone passes around the periphery of the flame and is not completely oxidized, at least not at this point. The fate of this quench zone material depends upon the conditions it is subjected to downstream from the flame. Turbulent mixing downstream may transport a portion of this partially oxidized mixture into the region where postflame reactions occur in the hot combustion products, as shown in Figure 7. Here these quench zone products may undergo further oxidation or pyrolysis depending upon the composition of the bulk stream at the point where mixing occurs.

Thus far only oxidative reactions have been discussed. The term pyrolysis refers to those reactions occurring either in the absence of oxygen or where the oxygen concentration is low enough that reactions between hydrocarbon species are more probable than reactions of hydrocarbons with oxygen or oxygen-containing species. Often in a combustion system most or all of the oxygen is consumed in the flame. The gases in the postflame region contain low concentrations of oxygen. Thus pyrolytic reactions are most likely to occur here. These reactions are simply mentioned in passing as part of the discussion of the fate of materials in the quench layer. They will be discussed in detail later when the nature of postflame reactions is considered.

2.4 IGNITION

The process of ignition is not specifically identified in Figures 3 or 5. Spatially, the ignition process occurs at the interface of the precombustion and combustion zones. Phenomenologically, ignition is the transition from a relatively low temperature state where the fuel is largely unoxidized to a condition where the temperature is sufficiently high that the rapid exothermic reactions typical of combustion are initiated and sustained. Chemically, ignition is the result of chain branching reactions. The concept of chain branching was introduced in Equation (16). These reactions produce a rapid increase in the local concentration of free radicals and other active species, which in turn react with the fuel molecules or fragments thereof. A fraction of them also diffuse upstream to initiate precombustion reactions.

The phenomenological definition presented above includes the proviso that combustion be sustained. A local portion of a fuel-air mixture may be ignited by the infusion of energy from some outside source, for example, a spark. If conditions are not correct and the combustion process is not sustained, that is if the flame or combustion wave that is actually created does not propagate through the remainder of the fuel mixture, then the ignition is said to be false. A number of factors may contribute to false ignition. The composition of the fuel-air mixture may be such that it is beyond the lean or the rich flammability limits. Also, too much diluent may be present. In either case, if the energy released is insufficient to maintain the temperature necessary for combustion to continue, then the initial flame kernal will not propagate through the remainder of the mixture. Another possibility is that the energy in the spark or other outside energy source is insufficient to insure that combustion is sustained even though the mixture ratio and other conditions are all favorable. The energy requirement of ignition is a complex subject, varying with mixture ratio (for example, the energy required increases rapidly as the lean flammability limit is approached) and with the time interval over which the energy is added to the system. For further discussion see Reference 4.

Practically, the ignition of a combustible mixture may be accomplished in a variety of ways. In the situation illustrated in Figures 3 and 5, once ignition has occurred it continues so long as there is a fresh supply of combustible mixture flowing into the tube. Two common methods of actually initiating the process are an electrical spark or a hot spot. The reader will recognize examples of these as the spark plug in an automobile engine and the glow plug in a model airplane engine, respectively. Ignition may be produced by other means, such as the rapid compression of a gas or gas-liquid system which in turn raises its temperature to the ignition point. This, of course, is what happens in a compression ignition (Diesel) engine. In some situations (particularly when the fuel is a solid or a liquid adsorbed on a solid) local chemical reactions within an aggregate of the fuel may raise its temperature to the ignition point. The result is usually referred to as spontaneous combustion or ignition.

Finally, it should be noted that ignition of gaseous, liquid and solid fuels (limited to mono- and bipropellants in the latter two cases) may be initiated by the propagation of a high velocity pressure wave through the fuel air mixture. Locally high temperatures are produced as the pressure wave passes through the mixture. This process is usually referred to as detonation. Practical examples are propellants and explosive materials.

2.5 COMBUSTION

The mechanisms by which hydrocarbons oxidize at the high temperatures encountered in a flame differ considerably from those just discussed, which occur at the much lower temperatures associated with precombustion reactions. Complex fuel molecules are not thermally stable at combustion temperatures. As fuel molecules enter the combustion region their internal energy increases rapidly due to collisions with very energetic high temperature species. Bond rupture occurs. This cracking or dissociation of fuel molecules produces simpler monatomic or diatomic species that react to form combustion products. Thus in the combustion region fuel molecules are oxidized to combustion products more directly than in the precombustion region where oxidation occurs via a series of steps as outlined in Equations (10) through (16).

In addition to CO_2, CO, H_2O and H_2, which are the predominant products of combustion, trace amounts of other species such as nitric oxide and particulate substances may form or begin to form in the combustion region. These trace substances are of special interest for they are among the major air pollutants produced by combustion.

Two basic approaches, thermodynamics and kinetics, are useful in the study of combustion. Thermodynamic analysis reveals information about the chemical and physical changes that can occur, but nothing about how rapidly these changes will occur. Conversely, kinetic analysis provides information on how quickly changes can occur but does not predict the extent of change that is ultimately possible.

It was noted earlier that the bulk of the chemical potential energy stored in the bonds of the fuel molecules is converted into thermal energy within the combustion region. The energy liberated when a fuel molecule combines with a theoretically correct quantity of oxygen to yield complete oxidation products is referred to as the heat of combustion of that fuel. For example:

$$C + O_2 \rightarrow CO_2 + 94.0 \qquad \frac{kcal}{mol\ C} \qquad (17)$$

$$CH_4 + O_2 \rightarrow CO_2 + H_2O + 212.8 \ \frac{kcal}{mol\ CH_4} \qquad (18)$$

$$\tfrac{1}{2}H_2 + O_2 \rightarrow \tfrac{1}{2}H_2O + 68.3 \qquad \frac{kcal}{mol\ H_2} \qquad (19)$$

The heat of combustion $(-\Delta H_C)$ is written with a negative sign. This is a thermodynamic convention and indicates that energy flows out of the system (in the case of combustion the system is usually defined to be the reacting gases). The heat of combustion of any fuel at standard conditions can be calculated from the standard heats of formation of the fuel and of the oxidation products.

$$\Delta H_C = \sum_{j=1}^{m} H_{f,j} - \sum_{i=1}^{p} H_{f,i} \qquad (20)$$

ΔH_C is the heat of combustion at standard conditions and $H_{f,i}$'s are the heats of formation of the fuel molecule(s) from their elements, solid carbon and molecular hydrogen. i is an index on fuel species. Similarly $H_{f,j}$'s are the heats of formation at standard conditions of the product molecules and j is an index on product species.

For purposes of performing a thermodynamic analysis of a combustion system, it is often assumed that the chemical reactions take place so rapidly that there is no time for exchange of heat between the reacting gases and their surroundings. This assumption is not entirely correct. Even at high temperatures characteristic of combustion, chemical reactions are not instantaneous. Reaction rate data on high temperature reactions of interest are given in References 9, 10 and 11. If the assumption of instantaneous reaction is accepted, the reaction will take place adiabatically, that is, without any transfer of heat to the surroundings. All of the heat liberated during combustion is used to break chemical bonds and to heat the reacting mixture. The maximum or peak temperature, the pressure and the composition of the reacting gases at this temperature and pressure depend upon the nature of the combustion process itself. For purposes of discussion here, only two paths will be considered: first, the case when the volume of the reacting gases is constant, and second when the pressure of the reacting gases remains constant. Combustion in a closed vessel is an example of constant volume combustion. Combustion of a gas flowing in a pipe as shown earlier in Figure 5 is an example of constant pressure combustion.

It is shown in thermodynamic texts that internal energy, U, does not change during an adiabatic constant volume process, *i.e.*,

$$\Delta U = 0 = \sum_{j=1}^{m} n_j U_j T_f - \sum_{i=1}^{p} n_i U_i T_o \tag{21}$$

where
   $n =$ mole fraction
   $U =$ internal energy
   $T_o =$ initial temperature (absolute units)
   $T_f =$ adiabatic flame temperature (also absolute units)
   $i =$ index denoting reacting species (fuel, oxygen)
   $j =$ index denoting product species (*i.e.*, equilibrium species distribution at peak flame temperatures).

Generally, the initial temperature, pressure, and composition of a reacting mixture are known. All other quantities, the final temperature, pressure and composition are unknown. Note that "final" here means the peak or adiabatic flame condition and not the conditions in the effluent. If the product distribution has a total of m species, there will be a total of $m + 1$ unknowns. The pressure and temperature contribute one unknown each; there are $(m - 1)$ unknowns with respect to composition since the composition of the mth species is fixed once the total system composition is known. Thus calculation of the final pressure, temperature and composition requires m equations in addition to Equation (21). An equation of state such as the ideal gas law can be utilized along with $(m-1)$ linearly independent equilibrium equations for the m assumed equilibrium species. Since the final composition depends upon both the final temperature and pressure, both of which are unknown, an iterative process is required to solve the system of $(m+1)$ independent equations. This is best done with the aid of a high speed digital computer.

A constant pressure adiabatic process is one in which the enthalpy does not change. The equation analogous to (21) is

$$\Delta H = 0 = \sum_{j=1}^{m} n_j H_j T_f - \sum_{i=1}^{p} n_i H_i T_o \tag{22}$$

where the notation is identical with the exception that H denotes enthalpy. In this case the final pressure is known (it is identical to the initial pressure). However final temperature and composition are unknown. These are obtained by simultaneous solution of a set of m linear independent equations.

No detailed calculations of adiabatic flame temperature and equilibrium flame composition will be presented here. The reader who is interested may refer to Reference 13. This discussion will be limited to an examination of the results of this study and to some general conclusions that may be drawn from this type of thermodynamic analysis.

Figure 8 presents the results for constant-volume adiabatic combustion of n-octane. Adiabatic flame temperature and the equilibrium product distribution for 15 species are shown as a function of equivalence ratio. The definition of equivalence ratio is based on air-fuel ratios (see Equation 8). This is readily verified by observing the variation in the species CO, H_2 and O_2 with Ø. The concentration of oxygen is greatest when Ø is greater than one, which corresponds to lean combustion as it should. The converse is true for CO and H_2.

Note first of all that this thermodynamic analysis confirms an earlier statement to the effect that complex reactant molecules, such as the octane fuel C_8H_{18}, undergo wholesale bond rupture to produce rather simple species. With the exception of the triatomic species CO_2 and H_2O, which are relatively stable products of combustion, all other species with a structure more complex than diatomics, i.e., HC•, NO_2,

$$\overset{\parallel}{\underset{O}{}}$$

HCH and CH_4, have concentrations several orders of magnitude lower

$$\overset{\parallel}{\underset{O}{}}$$

than the diatomic species. When formulating a set of equations for a thermodynamic analysis it is necessary to include all species that might

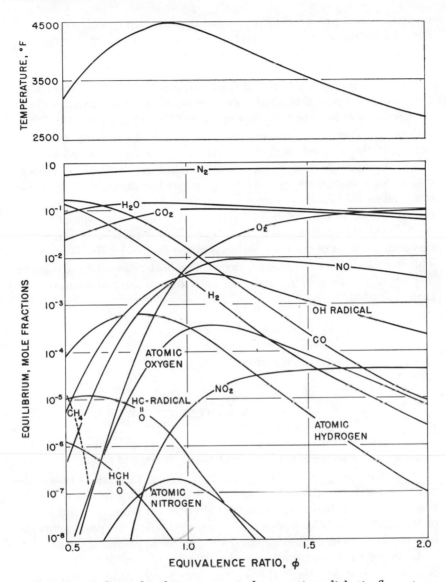

Figure 8. Relationship between equivalence ratio, adiabatic flame temperature and equilibrium product distribution for the constant volume adiabatic combustion of *n*-octane. Initial temperature 77°F and initial pressure 10 atm. Adapted from tables in Reference 12. Equivalence ratio is computed from air-fuel ratios (Equation 8).

be present at equilibrium. The results of the analysis may reveal that the concentrations of some species assumed to be present are so small as to be inconsequential. Thermodynamics cannot, however, do the reverse, that is, predict the concentration or occurrence of any species not initially postulated to be present in the final product distribution. Equilibrium flame compositions for a number of other cases of interest will be found in References 13 and 14.

A second observation of particular interest is that the maximum adiabatic flame temperature occurs near the stoichiometric mixture. There are a number of reasons for this. Excess fuel ($\emptyset < 1$) or excess air ($\emptyset > 1$) acts as a thermal sink and lowers the adiabatic flame temperature. A portion of the heat released by combustion must be expended to raise these excess components to the flame temperature. Increasing the fraction of inert substances such as nitrogen has the same effect on flame temperature. When the mixture is rich ($\emptyset < 1$), there is an additional consideration. When excess fuel is present, rather large quantities of partial oxidation products such as CO and H_2 are produced. Whenever this occurs not all of the energy stored in the fuel molecules is released. This is easily understood by observing the heat released by oxidation of carbon first to carbon monoxide and then of carbon monoxide to carbon dioxide.

$$C + \tfrac{1}{2} O_2 \rightarrow CO + 26.4 \ \frac{kcal}{mol} \tag{23}$$

$$CO + \tfrac{1}{2} O_2 \rightarrow CO_2 + 67.4 \ \frac{kcal}{mol} \tag{24}$$

$$+ \ \overline{\phantom{CO + \tfrac{1}{2} O_2 \rightarrow CO_2 + 67.4 \ \frac{kcal}{mol}}}$$

$$C + O_2 \rightarrow CO_2 + 93.8 \ \frac{kcal}{mol} \tag{25}$$

Recall that the heat of combustion is defined as the energy released when a fuel molecule is combined with a theoretically correct quantity of oxygen to completely oxidize the fuel. When the fuel is only partially oxidized, the entire heat of combustion is not released; some of this still remains bound up as potential chemical energy in the bonds of the partial oxidation species. Therefore, the final or peak temperature produced will be lower than if the fuel were completely oxidized.

Adiabatic flame temperature is of particular interest here for, as will be seen later, it influences the formation of nitric oxide, which is one of the major pollutants generated by combustion processes. It depends upon a number of variables in addition to the equivalence ratio. For example, the higher the initial temperature, T_o, the greater the adiabatic flame temperature. This is apparent from examining Equations (21) and (22) for constant volume and constant pressure combustion, respectively.

Fuel composition influences the adiabatic flame temperature also. This is illustrated in Figure 9. The information in this figure was taken

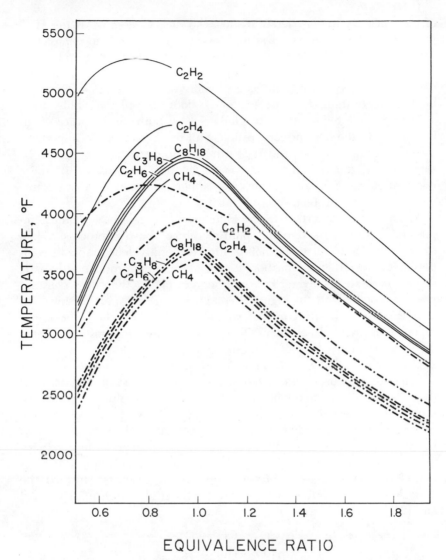

Figure 9. Relationship between fuel composition and adiabatic flame temperature. Solid curves are for constant volume adiabatic combustion and broken curves are for constant pressure adiabatic combustion. Initial pressure 10 atm. in all cases. Equivalence ratio is computed from air-fuel ratios. Adapted from tables in Reference 12.

from the same source as that in Figure 8. As the H/C ratio of the fuel decreases, or as y in the expression CH_y decreases, the adiabatic flame temperature increases. This is related to a number of factors, including the heats of combustion of the different fuel molecules and the quantity of air necessary to completely oxidize the fuel. Generally, as the H/C ratio of the fuel decreases less and less air is necessary to completely oxidize the fuel. This tends to increase adiabatic flame temperature by reducing the quantity of inert nitrogen that must be present. It should, of course, be recognized that as fuel composition changes so does the heat of combustion. The quantity of air does not tell the whole story; adiabatic flame temperature is influenced by both the quantity of air required and the heat of combustion. Still other factors are important. The percentages of CO_2 and H_2O produced depend upon y, as can be seen from Figure 1. This is important because dissociation of these combustion products has an important influence upon adiabatic flame temperature, as will be discussed shortly.

Another very important factor influencing the adiabatic flame temperature is the thermodynamic path of the combustion process. When the combustion of identical fuels at identical equivalence ratios is compared, it is seen that combustion at constant volume produces adiabatic flame temperatures several hundred degrees higher than does constant pressure combustion. In Figure 9 combustion at constant pressure, 10 atm, is compared with constant volume combustion where the initial pressure also is 10 atmospheres. The final pressures for constant volume combustion are necessarily greater.

The higher adiabatic flame temperature associated with constant volume combustion is associated with two factors. First, a portion of the difference is due to the fact that during constant volume combustion the reacting gases do not expand and no pdV work is done on the surroundings. Thus, additional energy remains within the system as internal energy and is manifested as higher adiabatic flame temperatures. Dissociation is also an important factor. From Figure 8 it is apparent that at flame temperature equilibrium is shifted toward the dissociative side, and therefore toward the right in Equations (26) and (27).

$$CO_2 \rightarrow CO^\bullet + O^\bullet \qquad (26)$$
$$H_2O \rightarrow HO^\bullet + H^\bullet \qquad (27)$$

The degree to which dissociation occurs depends upon both the pressure and the temperature. The Law of Mass Action (known also as LeChatelier's Principle) predicts that as the total pressure is increased, dissociation will decrease because the volume of the associated species is less than the volume occupied by the dissociation products. Conversely, an increase in temperature increases the degree of dissociation at equilibrium. The effects of pressure and temperature are not independent of one another during constant volume combustion. The increase in pressure causes a decrease in dissociation. Since dissociation is an endothermic process, a decrease in dissociation produces an increase in adiabatic flame temperature, which in turn tends to increase dissociation. The net result is a complicated equilibrium involving both temperature and pressure.

The relationship between pressure, degree of dissociation and flame temperature can be better understood by examining Figure 10. The adiabatic flame temperature produced by the combustion of *n*-octane increases slightly as the system pressure is increased from 1 to 100 atmospheres. This coincides to an overall decrease in dissociation, which is apparent from comparisons of the variation in concentrations of the species H_2O, CO_2, CO, $O^{\bullet}$ and $H^{\bullet}$. As the combustion pressure increases, the concentration of H_2O and CO_2 increases while that of the dissociation products CO, $O^{\bullet}$ and $H^{\bullet}$ decreases. Thus as pressure is increased, the equilibrium in Equations (20) and (27) is shifted toward the left, as LeChatelier's Principle predicts.

It has been noted that at the high temperatures attained during combustion only relatively simple atomic, diatomic and free radical species are stable. This does not imply that more complex species cannot exist in a combustion environment. They can and often do but when this occurs it represents a significant departure from equilibrium. Discussion of nonequilibrium processes in flames is an important subject that will be taken up shortly. For the moment the focus will be on equilibrium processes.

Mechanistically the elementary reactions that occur in flames usually involve two and sometimes three body collisions. The probability that any two species will collide and react is much greater than the probability that three species will collide and react. Consider the reaction of the radicals $H^{\bullet}$ and $OH^{\bullet}$. Collision of these two species will produce the water molecule.

$$H^{\bullet} + {}^{\bullet}OH \rightarrow H \cdots O \cdots H^{*} \tag{28}$$

The water molecule bears an asterisk (*) to indicate that it is produced in an excited state. It results from the collision of two energetic species, and the product $H \cdots O \cdots H^{*}$ contains too much internal energy to be thermodynamically stable. Unless collision occurs with some third body M, which can accept some of this excess energy so that

$$H \cdots O \cdots H^{*} + M \rightarrow H_2O + M \tag{29}$$

dissociation will occur.

$$H \cdots O \cdots H^{*} \rightarrow HO^{\bullet} + H^{\bullet} \tag{30}$$

$$\text{or } H \cdots O \cdots H^{*} \rightarrow H^{\bullet} + {}^{\bullet}OH \tag{31}$$

The half life of the intermediate species $H \cdots O \cdots H^{*}$ is on the

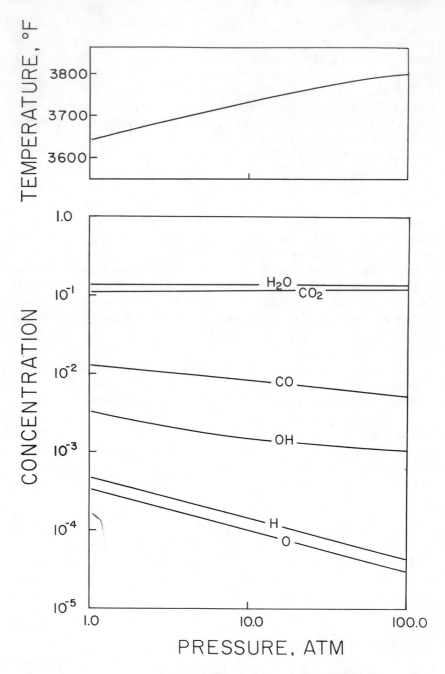

Figure 10. Adiabatic flame temperature and flame concentrations of selected species for combustion of *n*-octane at different pressures. Equivalence
ratio is computed from air-fuel ratios. Adapted from tables in Reference 12.

order of 10^{-13} seconds. This is several orders of magnitude less than the average time interval between successive two body-collisions in gases. Therefore unimolecular dissociation of the species $H \cdots O \cdots H^*$ via Reactions (30) or (31) is much more probable than stabilization via Reaction (29).

Equilibrium, when applied to combustion, denotes a condition wherein the rates of the many reactions occurring are rapid enough to maintain the concentration of each species in equilibrium with all of the other species present regardless of how rapidly the temperature and pressure of the system may be changing. Equilibrium is a dynamic condition from a mechanistic viewpoint. Reactions, particularly those resulting from two body collisions, are continually occurring even though the net result of all these reactions is no overall change in the system composition.

Equilibrium is not necessarily achieved during combustion. Reactions in the precombustion region are usually not rapid enough for equilibrium to occur. The same is true for some reactions in the flame itself and for many of the reactions in the postflame zone. In fact, high energy species in the flame and postflame regions may not even be equilibrated with respect to internal partitioning of energy between their own internal vibrational, rotational and electronic energy levels. These "hot" or "chemiionized" species may emit electromagnetic radiation in the visible region and are responsible, at least in part, for the characteristic luminosity of flame.

Figure 11 illustrates the partitioning of the energy released during the combustion process. The greatest part of this energy resides in the product molecules themselves. At high temperatures the energy levels within combustion product molecules may not be equilibrated, at least not initially. Excess energy in the electronic energy levels may be lost as electromagnetic radiation, the visible portion of which is called light. At the same time the energy not radiated is distributed among the other internal energy levels of the molecules and exchanged between molecules as collisions occur. These processes of exchange of energy between molecules are called thermalization.

Ultimately the energy residing in the product molecules is either converted into work by expansion of the working fluid or simply remains as heat energy to be degraded by transfer from the working fluid to the environment and by mixing of the working fluid with the environment. Classical Carnot cycle analysis can be used to determine the maximum amount of work that can be obtained from a combustion process. Consider a hypothetical combustion process in which the heat of combustion is released instantaneously at temperature T_c. If heat is rejected from the process at temperature T_e (corresponding to the

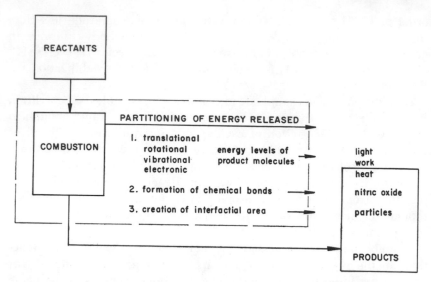

Figure 11. Partitioning of the energy released during combustion (see Reference 15).

effluent temperature), then the maximum amount of work that can be accomplished is

$$W = j\Delta H_c \frac{T_c - T_e}{T_c} \qquad (32)$$

where ΔH_c is the heat of combustion and j is a conversion factor, the mechanical equivalent of heat. The maximum thermal efficiency, η_t, of the process is

$$\eta_t = \frac{T_c - T_e}{T_c} \times 100 \qquad (33)$$

The actual thermal efficiency of any combustion process will be less than the theoretical efficiency because real systems are not reversible in the thermodynamic sense. The thermal efficiency is introduced at this point not only because it will be useful in later discussions but also to point out that it is a basically different quantity than combustion efficiency. Combustion efficiency is usually based on the ratio of the actual energy released in a combustion process, Q, to the heat of combustion.

$$\eta_c = \frac{Q}{\Delta H_c} \times 100 \qquad (34)$$

Typically, combustion efficiency is quite high, often approaching 100%.

Thermal efficiencies for most combustion processes encountered are much lower, almost always below 50 or 60%.

One reason why less than the total available energy may be released during combustion was discussed earlier, and it was noted that whenever the mixture is rich there is insufficient oxygen to completely oxidize the fuel component. A portion of the heat of combustion remains as chemical potential energy in the bonds of the partial oxidation products. In essence the partial oxidation products may be envisioned as fuels themselves and Equation (20) may be written to represent the "heat of combustion" for their own complete oxidation.

There are other reasons for the incomplete release of available energy, which may apply to lean or stoichiometric as well as rich combustion. Nonequilibrium processes arising as a result of kinetic limitations play a role in this. It has already been noted that dissociation at high temperatures is an endothermic process. Not only does this reduce the thermal efficiency by reducing T_c in Equation (33), but if, as the combustion products cool, the reverse process (association) does not occur because of kinetic limitations then species like carbon monoxide appear in the effluent even though stoichiometric considerations predict no carbon monoxide at overall stoichiometric or lean mixture ratios (refer to Figures 1 and 2). Thus, a portion of the available energy is not released because it is consumed by the endothermic formation of energy-rich molecules. The formation of NO and of carbonaceous particles are other examples of species whose formation prevents the complete conversion of the heat of combustion to either heat or work. In the case of carbonaceous particle formation not only does a portion of the available energy reside in the particles themselves, for carbon is a fuel in its own right as seen from Equation (17), but the energy also must be partitioned to create the large interfacial areas associated with small particles. The production of particles by combustion processes is a subject that will be dealt with in much greater detail later. It is of great interest because combustion processes are often sources of large quantities of very small particles known as combustion nuclei. These particles, which are often less than 0.01μ in diameter and have very large amounts of interfacial surface per unit mass, play an important role both physically and chemically in the atmosphere and are considered to be among the major air contaminants emitted by combustion processes.

They are also of interest because of their influence on partitioning of energy in the combustion process.[16] Very small particles can remain in thermal and dynamic equilibrium with the working fluid as it passes through the flame zone.[17] As a result they become incandescent and

emit a continuum of electromagnetic radiation.[18] This continuum is
described by the Steffan–Boltzman law for black body emitters. The
relative intensity of different wavelengths depends upon the absolute
temperature of the particle. In Figure 12 radiant energy emitted by

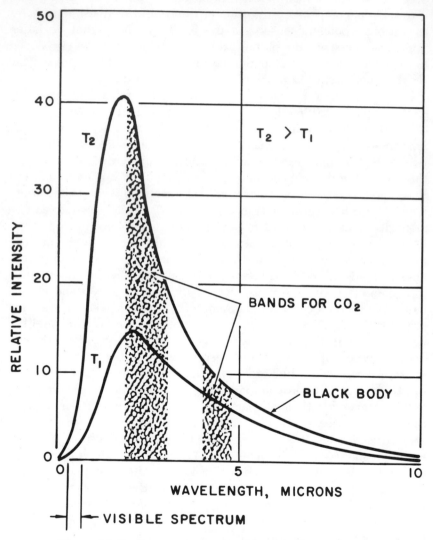

Figure 12. Radiant energy emitted by particles and gases.

particles is compared with the banded radiation emitted by gases.
Gaseous species such as the triatomics CO_2 and H_2O and various ionic
or free radical species emit discrete wavelengths (or bands) of radiation

as they undergo transition from higher to lower energy states. These bands are depicted as shaded areas. The negatively skewed solid curve represents the continuum of energy emitted by particles. The area under the curve is proportional to the total energy emitted. Therefore particles may result in the emission of very appreciable quantities of energy.[19] In fact, a major portion of the luminosity of many flames can be attributed to particles. In addition, when the quantity of electromagnetic radiation emitted from a flame, due to particle emission or otherwise, becomes appreciable, the combustion process can no longer be treated as an adiabatic system. Peak flame temperatures are significantly less than the theoretical adiabatic temperature.

2.6 NITROGEN OXIDES

Nitric oxide is one of the principal contaminants emitted by combustion processes. The formation of this compound is a complex process that involves several elementary chemical reactions. It involves both equilibrium and nonequilibrium processes that take place in the precombustion, combustion and postflame regions.

The following reactions

$$N_2 + O^\bullet \rightleftharpoons NO + N^\bullet - 75 \frac{kcal}{mol} \tag{35}$$

$$N^\bullet + O_2 \rightleftharpoons NO + O^\bullet + 31.8 \frac{kcal}{mol} \tag{36}$$

are referred to as the Zeldovich Mechanism for nitrogen fixation. The chain aspect of this mechanism for the conversion of molecular nitrogen and oxygen to nitric oxide is illustrated in Figure 13.

It is not generally possible to study the behavior of a species by isolating a few reactions in which it participates from the remainder of the system. The formation of nitrogen oxides is coupled to numerous other reactions that occur simultaneously. For example, the following reaction is a major source of oxygen atoms.

$$O_2 + M \rightleftharpoons O^\bullet + O^\bullet + M \tag{37}$$

Oxygen atoms are reactants in Equation (35). They also react with a host of fuel fragments. Figuratively, there is competition for the available oxygen atoms between Reaction (35) and various oxidation reactions. The concentration of these oxygen atoms is determined by the relative rates of oxygen atom production via Reaction (37) and similar reactions and consumption of oxygen atoms via Reaction (35), etc. At the same time, the rate of Reaction (35) depends upon the oxygen atom concentration.

Reaction (38) is a major source of nitrogen atoms.

$$N_2 + M \rightleftharpoons N^{\bullet} + N^{\bullet} + M \tag{38}$$

Nitrogen atoms also participate in nitric oxide formation via Reaction (36).

The rates of each of the preceding reactions, (35) through (36), increase rapidly as the temperature increases. In the cases of Reactions (37) and (38) the activation energies are essentially the energies neces-

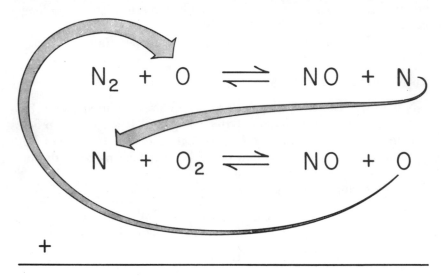

Figure 13. Chain mechanism for the fixation of nitric oxide.

sary to break the oxygen and nitrogen molecular bonds. These are 118 and 225 kcal/mole, respectively, and are extremely high values for activation energy. Therefore, it is not surprising that the quantity of nitric oxide formed increases rapidly as the flame temperature increases.

It is apparent, then, that the flame has two roles with respect to the formation of nitric oxide. First, it provides the thermal energy required to fix the nitrogen. Second, it provides the reactions initiating the chain reactions that produce nitric oxide.

In addition to the four reactions cited above, there may well be others that influence the formation kinetics of nitric oxide. For example, the exchange reaction shown in Equation (39) is very fast.[20]

$$N^\bullet + {}^\bullet OH \rightleftharpoons NO + H^\bullet + 39.4 \frac{kcal}{mol} \tag{39}$$

In addition in rich flames, reaction (40) may occur:[20]

$$CH^\bullet + N_2 \rightleftharpoons N^\bullet + HCN \tag{40}$$

This last reaction may be particularly important as the primary source of nitrogen atoms at lower temperatures where the rate of Reaction (38) would be low.

A number of investigators have predicted the quantities of NO experimentally observed in different types of flames. From these studies, it appears that NO formation in many combustion systems of practical interest is limited by formation kinetics.[21,22,23,24] This type of behavior is illustrated graphically in Figure 14. Consider a combustible mixture flowing in a pipe as was shown in Figure 5. As the mixture passes through the precombustion zone, a slight rise in temperature occurs, due primarily to counter-current convective transfer of heat from the combustion zone. Subsequently a rapid rise occurs in the combustion zone itself, followed by a gradual temperature decay in the postflame region. The locus of temperature in the flowing gas is shown in the upper portion of Figure 14. If the elementary reactions leading to the formation and decomposition of NO were sufficiently rapid, the nitric oxide would always be equilibrated with all other species in the reacting gases. In this case, the dotted line would represent the concentration of nitric oxide as a function of axial-position in the pipe. This line is marked "equilibrium" and the Zeldo-vich chain reactions are shown to indicate that equilibrium is assumed with respect to these reactions. Note that the maximum concentration of nitric oxide $[NO]_{t,max}$ coincides with the peak flame temperature achieved.

The actual locus of NO concentrations deviates from this equilibrium case. Some studies have indicated the early appearance of nitric oxide, establishing its occurrence in the flame front at concentrations well in excess of that expected from equilibrium considerations. The nitric oxide so formed has been termed *prompt NO*.[20] The mechanism of this phenomenon is not well understood and may involve the formation of NO via a reaction such as

$$N^\bullet + {}^\bullet OH \rightleftharpoons NO + H^\bullet \tag{41}$$

where the nitrogen atom appears at relatively low temperatures as a result of another reaction such as (40). The hydroxyl radicals required

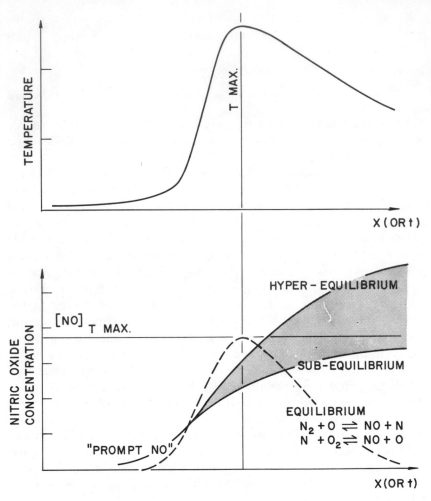

Figure 14. Formation of nitric oxide.

in Reaction (41) are common in combustion systems, particularly when water vapor is present.

Various kinetic analyses indicate that the rate of nitric oxide formation usually lags behind the equilibrium value well into the combustion zone. Apparently, the forward reaction rates of Reactions (35) and (36) are not rapid enough to produce an equilibrium concentration of NO.[21] Nitric oxide also may continue to form well into the postflame region. The concentration of NO ultimately produced may be less or greater than that which theoretically corresponds to equilibrium at the peak flame temperatures. This range of locii is denoted by the shaded region in Figure 14. Excess or hyperequilibrium quantities of NO, which can occur in certain types of flames, may possibly be explained by the production of abnormally large concentrations of $O^{\bullet}$ and/or $N^{\bullet}$ in such flames, which in turn both drive the Reactions (35) and (36) toward the right and increase the forward reaction rates. On the other hand, ultimate concentrations of nitric oxide less than that predicted for the maximum flame temperature can be due either to limitations on the formation kinetics as discussed previously or, where the combustion is rich, to reaction of a portion of the nitric oxide in the postflame region.[22] In rich combustion systems, species such as $H^{\bullet}$ may persist well into the postflame region, and a portion of the nitric oxide may react via reactions such as the reverse of (39). In the case of lean combustion, the rate of decomposition of nitric oxide in the post-flame region via the reverse of Reactions (35) and (36) is slow. Not only do the rates of the reverse reactions of (35) and (36) decrease rapidly as temperature decreases, but the concentrations of oxygen and nitro-gen atoms, which are reactants in the reverse reactions, also decrease rapidly.

When the formation of nitric oxide is limited by the formation ki-netics as portrayed in Figure 14, then the longer the time that high temperatures are sustained in a combustion system, the larger the quantity of nitric oxide that can be expected to form. This is true, at least to the point where the reacting mixture is equilibrated. Con-versely, since nitric oxide formation continues well into the postflame region, rapid quenching of the postflame gases by heat removal or by gas expansion will tend to reduce nitric oxide formation in combustion systems.

The formation of nitric oxide in combustion systems is for the most part determined by the peak temperatures achieved during combus-tion. In turn, the peak temperatures achieved depend upon a number of variables, including the mixture's stoichiometry, the quantity of diluents present, fuel composition, and the initial temperature of the fuel-air mixture. Figure 15 is a conceptual representation of the depen-dence of flame temperature on the equivalence ratio and the H/C ratio of the fuel. For any value of Ø and H/C the corresponding tem-perature is represented as a point on the three-dimensional surface shown.

Insomuch as nitric oxide is a pollutant, it is usually desirable to reduce its formation by exercising control over the above-mentioned process variables. Various known methods of reducing flame tempera-ture and thereby controlling the quantity of nitric oxide formed during combustion will be explained.

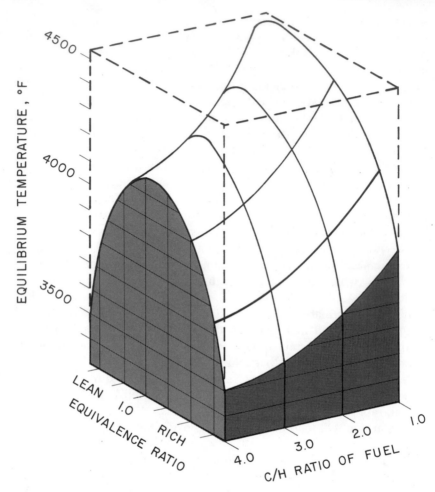

Figure 15. Relationship between equivalence ratio, C/H ratio of fuel, and flame temperature.

Flame temperature depends upon the stoichiometry of the fuel-air mixture. As shown in Figure 15 it is maximum near the stoichiometric ratio and decreases as the mixture becomes progressively richer or leaner for reasons previously discussed. The general type of relationship existing between nitric oxide production in a premixed flame and the equivalence ratio of the mixture is illustrated in Figure 16. Flame temperature is a maximum at or near the stoichiometric fuel-air mixture, while maximum nitric oxide formation usually occurs on the lean side of the stoichiometric point. This latter phenomenon is caused by the dependence of the rate of nitric oxide formation upon

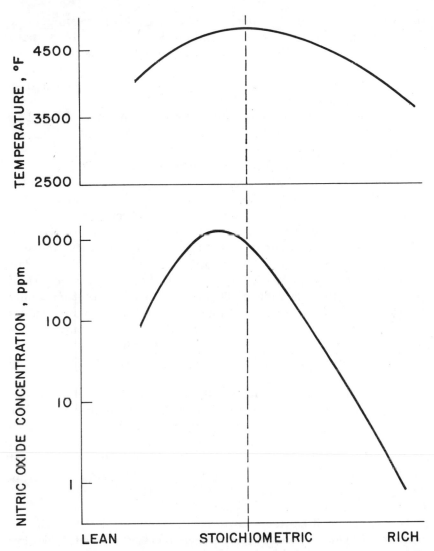

Figure 16. Relationship of flame temperature and equilibrium nitric oxide concentration.

both the temperature and oxygen atom concentration. Though peak temperatures occur near $\emptyset = 1$, oxygen atom concentrations increase as the mixture becomes progressively leaner, as can be seen by referring back to Figure 8. The net result of the rate dependency on both temperature and of oxygen atom concentration is a maximum in nitric oxide formation at a slightly lean mixture ratio. There is a distinct similarity between the shapes of the curves relating nitric oxide concentration to equivalence ratio in Figures 8 and 16. Actually these figures portray different quantities. Figure 8 depicts the relationship between Ø and [NO] at peak flame temperatures and Figure 16 illustrates the relationship between Ø and [NO] at the effluent from the combustion process.

One means of reducing nitric oxide is to purposely adjust the fuel-air mixture so that it is "off-stoichiometric" (*i.e.*, rich or lean). This approach utilizes excess air or fuel to cool the reacting mixture. There is, however, a limit to how rich or how lean a mixture can be and still maintain combustion. Combustion can be sustained only so long as the energy liberated equals or exceeds the energy required to raise the temperature of the fuel-air mixture to the ignition point. Mixtures within this range are flammable. The fuel-air mixture that corresponds to the maximum amount of air that can be present per unit of fuel and still ignite is called the lean-flammability limit of the mixture. The corresponding limit on the maximum amount of fuel that can be present per unit air is the rich flammability limit. Generally the flammability limits of a mixture are so skewed that considerably greater reductions in NO can be achieved through rich rather than lean combustion. This is implied by the curve in Figure 16. The extremities of the temperature and nitric oxide concentration curves correspond to the rich and lean flammability limits.

Combustion of rich mixtures results in a considerable reduction in the formation of NO. At the same time, large quantities of carbon monoxide and hydrogen are produced, as is shown by referring to Figures 1 and 2. Other partial oxidation products composed of fuel and fuel fragments are invariably produced. With the exception of hydrogen these species are also undesirable contaminants if they are simply exhausted to the atmosphere. Therefore, additional treatment of the combustion products is usually required.

This subject will be discussed in considerably greater detail in the next chapter when secondary combustion processes are considered. For the moment it is sufficient to note that an arrangement commonly referred to as *two-stage combustion* can be used. This arrangement is illustrated in Figure 17. The first stage is called the primary combus-

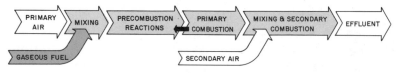

Figure 17. Two-stage combustion.

tion zone. The mixture entering this zone is rich, that is it contains excess fuel and the equivalence ratio $\emptyset_p$ in this zone is $\emptyset_p < 1.0$. Additional air (secondary air) is added after the primary combustion stage to facilitate the complete oxidation of products (CO, H_2, etc.) formed in the primary combustion zone. Enough secondary air is usually added to insure that the overall equivalence ratio for the process $\emptyset$ is considerably greater than 1.0. This provides for excess oxygen in the effluent and thereby increases the probability of oxidizing any partial combustion products from the primary combustion zone. This is important when secondary mixing is not complete, which is usually the case. Peak temperatures achieved in either zone for two-stage combustion are lower than those attained with single-stage combustion. The overall combustion process is no longer adiabatic whenever it is divided into two separate events for heat loss inevitably occurs between the two. The process in the second stage is called secondary combustion. Whether or not it is combustion in the classical sense or simply a high temperature, homogeneous gas phase reaction depends upon a number of factors including temperature, the concentrations of combustibles and of oxygen, the quantity of diluents present, and the nature of the mixing of first stage effluent with the secondary air.

A related but different method of reducing combustion temperature is to introduce inert substances. These substances absorb a portion of the energy released as they are heated from their initial temperature to flame temperature. At the same time little or no additional energy is released as a result of their presence. The net result is a reduced flame temperature. One method employed is recycling a portion of the effluent. This technique is indicated schematically by the upper dotted path in Figure 18. The recycle stream consists principally of N_2, H_2O

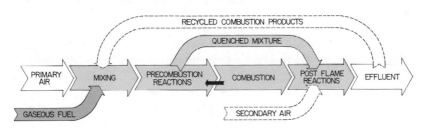

Figure 18. Two-stage combustion with recycle.

and CO_2. All of these absorb significant amounts of energy as they are reheated to the flame temperature. In addition to this absorption, a portion of the CO_2 and H_2O dissociate at flame temperatures as noted earlier in Equations (26) and (27). Dissociation is an endothermic

process and consequently can assist in reducing the flame temperatures.

The amount of combustion products that are recyclable is intrinsi-
cally limited by flammability considerations. The flammability envelope
for a combustible mixture is depicted in Figure 19. The ordinate cor-

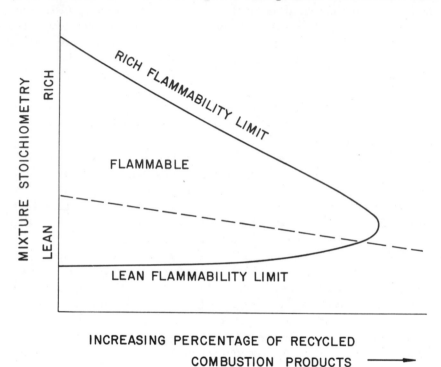

Figure 19. Flammability envelope for a combustible mixture.

responds to the limits for combustion of fuel in air. The rich and lean
flammability limits correspond to the points where the envelope intersects the ordinate. When the fuel-air mixture is diluted by the addition
of gaseous diluents, the rich and lean flammability limits approach one
another and finally converge. This is not unexpected since the larger
the quantity of diluent present, the smaller the excess of air or
fuel that will increase the sensible heat capacity of the mixture to the
point where the ignition temperature is no longer attained or sustained.
The dashed line is the locus of the path corresponding to progressive
dilution of a fuel-air mixture with combustion products. It is impossible to dilute a mixture beyond the point where the flammability
envelope is intersected and still sustain combustion.

As an alternative to recycling exhaust products, low pressure steam or liquid water can be added to the fuel-air mixture. In addition to absorbing energy as sensible heat and to dissociating at flame temperatures, the introduction of water as a liquid has an additional benefit in reducing combustion temperatures. Water has a very high latent heat of vaporization, and a large quantity of energy is absorbed in transforming it from the liquid to vapor phase.

The combined effect of mixture stoichiometry, recirculation of combustion products and of initial mixture temperature is displayed in Figure 20. Each surface illustrates how the concentration of nitric oxide produced varies as the equivalence ratio and quantity of exhaust products is varied continuously. The different surfaces correspond to finite increments in the value of initial mixture temperature such that $T_1 > T_2 > T_3$.

Finally, nitric oxide production generally increases as the combustion intensity (energy released/unit volume/unit time) increases. An explanation for this is that the volumetric heat loss does not increase proportionally, which results in higher flame temperatures and greater nitric oxide formation.

2.7 OXIDATION OF SULFUR

Fuels sometimes contain sulfur, which may be inorganically or organically bound. Inorganic sulfur (sulfides, etc.) generally occurs in solid rather than in liquid or gaseous fuels. Hence, discussion of this case will be deferred to Chapter 4. Organic sulfur compounds, i.e., ones in which carbon-sulfur bonds exist, may be present in liquid and gaseous fuels. Some compounds, for example, carbon disulfide (CS_2) and hydrogen sulfide (H_2S), could be used as fuels themselves. They rarely are, of course, because of the large concentrations of sulfur dioxide that would result. More commonly sulfur-containing compounds such as the ones mentioned above and others, notably the mercaptans, are added in small concentrations to common gaseous fuels. These trace sulfur compounds all have very powerful and distinct odors and they serve as odorants to warn of the presence of leaking gaseous fuels that would otherwise remain undetected save for unexpected explosions.

Whatever the source of sulfur in fuels, when the fuel is oxidized, the sulfur does likewise. The high temperature equilibrium shown in Figure 8 does not predict the presence of sulfur or any of its oxides. This is no great surprise since the presence of sulfur in the fuel was not considered in the computation. When sulfur is considered, equilibrium computations predict that at high temperatures the stable oxide of

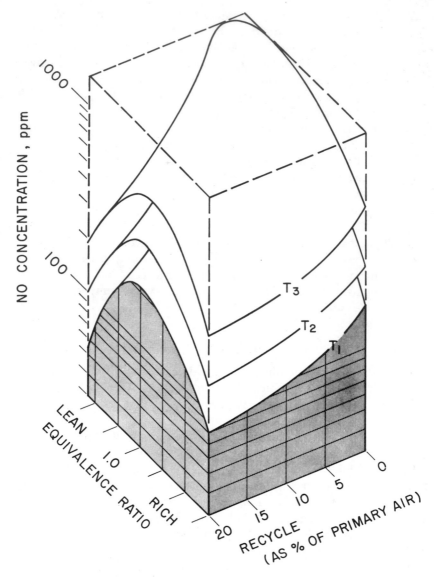

Figure 20. Nitric oxide production as a function of equivalence ratio, exhaust recycle and initial mixture temperature. Equivalence ratios computed from air-fuel ratios.

sulfur is sulfur dioxide (SO_2). Relatively little sulfur trioxide (SO_3), which is the more stable oxide at lower temperatures, will be formed. Discussion of the elementary reactions of sulfur in flames can be found in References 25 and 26.

2.8 POSTFLAME REACTIONS

The term *postflame* refers to that region located downstream of a combustion zone. It is a possible misnomer in that it includes a region that may emit luminous radiation, and therefore it is usually thought of as a part of the flame. The preponderance of this luminosity can generally be ascribed to particles that have been heated to incandescence in the combustion zone and continue to radiate in the visible band as they cool or continue to react in the postflame gases. This condition is particularly common in the case of rich combustion (consult Figure 4). The combustion zone itself is a relatively narrow blue zone near the neck of the Bunsen burner. However, when the fuel-air mixture is rich, the flame has a long yellow tail produced by incandescent carbon particles.

Many physical as well as chemical processes can occur in a postflame region. The reactants entering the region include particulate as well as gaseous species. The latter may be atomic, molecular, free radical and even ionic in character. These reactants participate in heterogeneous as well as homogeneous reactions, which in either case can be oxidative or pyrolytic in nature. There is no simple way to group or characterize the diverse reactions and processes taking place in this region. They will be discussed in the following order: radical recombination, CO oxidation, pyrolytic reactions, particulate formation and reactions thereof.

The phenomena occurring in the postflame region are particularly important for their strong influence on the quantities of many minor species present in the effluent.

Radical Recombination

It was noted earlier that the most probable type of reactions that radical and atomic species undergo in high temperature combustion systems are those resulting from two-body collisions. Generally the product species of such a collision contains too much energy for it to be stable and it dissociates. The products of such reactions are transitory.

Radical recombination reactions (alternatively referred to as chain termination reactions) such as

$$O^{\bullet} + O^{\bullet} + M \rightarrow O_2 + M \tag{42}$$

$$H^{\bullet} + {}^{\bullet}OH + M \rightarrow H_2O + M \tag{43}$$

require the simultaneous or nearly simultaneous collision of three bodies. The function of the third body, M, is to stabilize the product molecule by absorbing a portion of its internal energy. From a probabilistic point of view three-body collisions are very much less frequent than two-body collisions. Consequently rate of recombination of radicals formed during combustion is significantly slower than other high temperature reactions, and the products persist well into the postflame region.

In time recombination can occur. In the interim the hyperequilibrium concentrations of free radicals do influence other aspects of combustion. One example of this is their influence on nitric oxide formation. This was discussed in the previous section. Another aspect is the influence of hyperequilibrium concentrations of free radicals on the oxidation rate of carbon monoxide.

Oxidation of Carbon Monoxide

It is generally accepted[23, 24, 27, 28] that the most important reaction for oxidizing carbon monoxide to carbon dioxide in postflame gases is

$$CO + {}^{\bullet}OH \rightleftharpoons CO_2 + H^{\bullet} + 25 \frac{kcal}{mol} \tag{44}$$

The rate of this reaction is sufficiently rapid to keep the four species involved equilibrated as the postflame gases cool. The equilibrium constant for Reaction (44) is written in the usual manner as

$$K = \frac{[CO_2][H^{\bullet}]}{[CO][{}^{\bullet}OH]} \tag{45}$$

This can be rearranged as follows:

$$\frac{[CO]}{[CO_2]} = K^{-1} \frac{[H^{\bullet}]}{[{}^{\bullet}OH]} \tag{46}$$

As the postflame gases cool the concentrations of free radical species do not decrease as rapidly as expected from equilibrium considerations. This is due, of course, to the inherent slowness of recombination reactions. Kinetic analysis of the system of elementary reactions that are occurring in the postflame gas leads to the conclusion that the ratio $[H^{\bullet}]/[{}^{\bullet}OH]$ becomes very large as a result in particular of the hyperequilibrium concentration of atomic hydrogen. Since the kinetics of Reaction (44) are rapid enough, at least at higher temperatures, to keep the species CO, CO_2, ${}^{\bullet}H$ and ${}^{\bullet}OH$ equilibrated, the ratio of $[CO]/[CO_2]$ must be larger than otherwise expected to maintain the equality expressed in Equation (46).

At lower temperatures, after radical recombination has occurred, the shift Reaction (47)

$$H_2O + CO \rightleftharpoons H_2 + CO_2 \tag{47}$$

becomes the important reaction for equilibrating the amounts of CO and CO_2 in the effluent. However, since an excess of CO is established early during the cooling process and since the rate of the forward and reverse reactions in Equation (47) decreases with temperature, the excess CO is essentially "frozen" into the effluent products. In the case of rich combustion the explanation for CO in the effluent is obvious. The coupling between recombination reactions and CO oxidation, which is discussed above, is, however, one explanation for the simultaneous appearance of CO and oxygen at equivalence ratios slightly on the rich side of stoichiometric. This type of behavior is observed experimentally and illustrated in Figure 21. It is in contrast with Figures 1 and 2, which indicate that based on the stoichiometric considerations alone the appearance of carbon monoxide and oxygen should be a mutually exclusive event.

One interesting conclusion that can be gleaned from the above analysis is that the initial cooling of the postflame gases should be slow enough to allow for equilibration of all species, including radicals which are recombining, in order to minimize CO formation in combustion systems. This raises a perplexing situation, for this procedure is diametrically opposite to that proposed earlier to minimize nitric oxide production. Rapid cooling is favorable to low nitric oxide production and unfavorable to reducing carbon monoxide, and vice versa for slow cooling.[13] Thus, simultaneous reduction of CO and NO via positive control of postflame gas cooling rates appears to be, at best, a matter of compromise.

Pyrolysis

The term pyrolysis may refer to a variety of reactions. As the term is used here it means those reactions of hydrocarbons that occur in the absence or near absence of oxidizing species. In many cases reactions are endothermic and the necessary activation energy is supplied thermally. A number of pyrolytic reactions important in combustion systems will be discussed. These include the reaction of fuel species or products thereof with other hydrocarbon species, the dehydrogenation of hydrocarbons to produce species with greater unsaturation, and the cracking of hydrocarbons.

The nature of molecular reactions depends upon many factors, including the temperature and chemical nature of the environment. Elevated temperatures and a reducing (in contrast to an oxidizing) environment promote the occurrence of pyrolytic reactions. The postflame gases from all but relatively lean premixed flames have these conditions.

Pyrolytic reactions occur over a wide range of temperatures. At the higher temperatures bond rupture and dehydrogenation predominate in

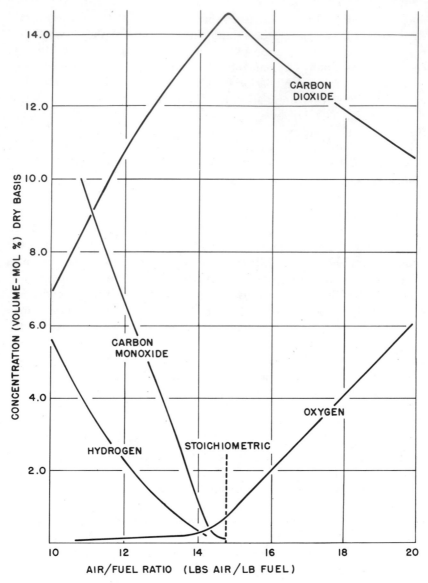

Figure 21. Experimental relationship between concentrations of effluent species.

pyrolysis. These endothermic reactions produce lower molecular weight species from more complex ones. At lower temperatures pyrolytic reactions are likely to involve the polymerization or association of species to produce higher molecular weight ones from smaller fragments.

A hypothetical system of reactions for pyrolysis of ethane is presented here.

$$C_2H_6 + M \rightarrow 2CH_3{}^{\bullet} + M \qquad (48)$$

$$2CH_3{}^{\bullet} + 2C_2H_6 \rightarrow 2CH_4 + 2C_2H_5{}^{\bullet} \qquad (49)$$

$$M + C_2H_5{}^{\bullet} \rightarrow H^{\bullet} + C_2H_4 + M \qquad (50)$$

$$H^{\bullet} + C_2H_6 \rightarrow H_2 + C_2H_5{}^{\bullet} \qquad (51)$$

$$M + 2C_2H_5{}^{\bullet} \rightarrow C_4H_{10} + M \qquad (52)$$

$$\overline{4C_2H_6 \rightarrow 2CH_4 + H_2 + C_2H_4 + C_4H_{10}} \qquad (53)$$

This hypothetical example illustrates several important points. First the initiation step, Reaction (48), has an activation energy of 88 kcal/mole, which is essentially the dissociation energy of the C—C bond, and therefore occurs only at relatively high temperatures. A second observation is that the remaining steps result in the production of a species that is less saturated (C_2H_4) than the reactants as well as species of both higher (C_4H_{10}) and lower (CH_4) molecular weight.

Similar reaction schemes may be written for other hydrocarbon species. Different hydrocarbons differ in ease by which they undergo pyrolysis. Some are more refractory than others and require higher temperatures to initiate the reactions, though hydrocarbons that are saturated initially can undergo lower temperature additions or poly- merization more readily.

Hydrocarbon species can enter the postflame region in at least two ways. First, they may originate in the flame, a particular likelihood for rich combustion. The hydrocarbon species from this source are likely to be relatively simple in structure and may be quite unsaturated due to dehydrogenation reactions that are possible energetically at high temperatures.

$$\begin{array}{ccccc} \text{Saturated} & \rightarrow & \text{Olefinic} & \rightarrow & \text{Acetylenic} \\ \text{Hydrocarbon} & \Delta & \text{Hydrocarbon} & \Delta & \text{Hydrocarbon} \end{array} \qquad (54)$$

In the case where the hydrocarbon is ethane, Reaction (54) would lead first to the olefin ethylene and finally to acetylene. A second major source of hydrocarbons was previously pointed out in the discussion of Figure 7. Turbulence in the postflame region can mix hydrocarbons from the quench layer with the hot postflame gases. The fate of these particular species depends upon the point at which they mix with the postflame gases, that is, whether the temperature is still so high as to cause bond rupture or whether they will participate in lower temper- ature reactions with other species already present.

An understanding of the nature and products of pyrolytic reactions is important because these reactions contribute to the distribution of hydrocarbon species in the effluent. There is yet another reason why the subject of pyrolysis is of concern. A *sequence* of pyrolytic reactions may lead to the *synthesis* of relatively large and complicated molecules in the effluent (see References 29, 30, 31, 32, 33). An example of this is illustrated in Figure 22. Two paths are shown for the synthesis of the

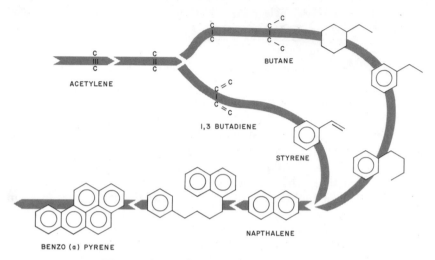

Figure 22. Pyrolytic synthesis of B (a) P.

polynuclear aromatic benzo-(a)-pyrene. As in the case of precombustion reactions, low temperature pyrolysis reactions proceed via a sequence of distinct intermediates. Another similarity between the two is that many parallel paths also exist, only two of which are suggested in Figure 22. Other things being equal, the tendency for hydrocarbons to form polynuclear species is:

$$Aromatics > Cycloolefins > Olefins > Paraffins \qquad (55)$$

A qualitative explanation for this rank order is found in Figure 22. First, unsaturation of the reactants promotes addition-type reactions between hydrocarbon species. Second, a ring structure is a convenient building block for more complex condensed ring structure.[34,35,36,37]

The discussion thus far has treated pyrolytic and oxidative reactions in the postflame region as if they were mutually exclusive. This is often found not to be the case. The postflame gases contain small quantities of oxygen. This is even found to be true on the rich side of stoichiometric (see Figure 21), much as CO can be found in the combustion products of lean mixtures. The reaction of hydrocarbons in this region can be viewed as a competition between oxidative and pyrolytic paths as illustrated in Figure 23. The greater the concentration of oxygen, the greater the rate, r_o, of oxidation. The addition of secondary air, as discussed earlier and shown in Figure 17, is an attempt to promote oxidation of any residual hydrocarbon species. Even when secondary air is added, the rate of pyrolysis, r_p, may be quite significant due to imperfect mixing of the secondary air with the postflame gases.

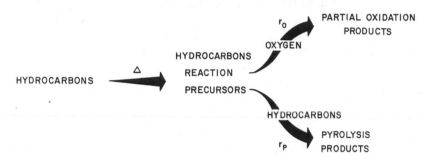

Figure 23. Competitive oxidative and pyrolytic paths for reaction of hydrocarbons.

Pyrolytic reactions are important for one additional reason. The products of these reactions may play an important role in the formation and growth of particulate matter.

Formation of Particulate Matter

The genesis of particulate matter in combustion processes is a subject of considerable interest. Much of the particulate matter emitted from combustion is very finely divided, and as such is considered to be an important air pollutant.

Basically particulate matter may originate from inorganic or organometallic substances inducted with either the fuel or the air from reactions of the fuel itself or from combinations thereof. Inorganic matter present in the fuel is called ash and cannot be destroyed by complete ignition of the fuel. The ash content is generally quite low for fuels that are premixed with the air. It is a much more important source of particulates when burning heavy liquid and solid fuels. These cases will be reserved for discussion later under the subject of diffusion flames. Organometallics may be present in the fuel, particularly as additive substances. These species are oxidized in the flame and the metal component appears as an inorganic oxide or salt in the postflame gases.

Finely divided inorganic material is present in the ambient air and is inducted into the combustion process. As it passes through the flame the high energy density environment may alter it physically as well as chemically. That is, its particle size distribution may be shifted toward smaller mean diameters as a result of vaporization and recondensation in the postflame gases. It will also most surely be oxidized to some degree if it is not already present in its highest oxidation state.

A final source of particulate matter is the fuel molecules themselves. It was pointed out in the previous section that those pyrolytic reactions occurring at lower temperatures can produce rather large molecules. As the molecular weight of these species increases the hydrogen-carbon ratio generally decreases. For example, the H/C ratio of benzo-(a)-pyrene in Figure 22 is 0.6, whereas this same ratio for the reactants is 1.0 or larger, depending upon which species is considered to be the precursor. The H/C ratio continues to decrease as the molecular weight of the species increases. Ultimately ratios of $0.00 < H/C < 0.50$ may be attained.

Actually, little is known about the exact mechanism of particulate growth from species with a score of carbon atoms to one with tens of thousands of carbon atoms. It is possible that the inorganic oxides mentioned above may act as nuclei for particles by absorbing higher molecular weight hydrocarbon species. Alternatively, as the organic molecules grow, their vapor pressure is decreasing at the same time that gas temperature is also decreasing. Organic particles may result from self-nucleation. Inorganic substances, such as halogens, which may be present in fuel additives can promote particle growth, perhaps by promoting dehydration of large organic molecules.

In addition to chemical mechanisms, particles may grow by processes that are essentially physical in nature. Agglomeration of small particles to produce larger aggregates stuck together in a rather open network of a few hundred Ångstroms have been observed. The particles consist of 10^3–10^4 crystallites, each composed of 5–20 sheets of carbon atoms with a length and breadth of 20–30 Å. Chain-like agglomerates from 100–2500 Å may ultimately form.

The growth of an agglomerate is illustrated diagrammatically in Figure 24. The individual sheets of carbon atoms are formed by extending the structure shown in the upper right-hand corner. The sheets combine to form crystallites as contained in the shaded area on the left. Finally the crystallites agglomerate in more or less random fashion to form larger aggregates with interstitial regions and areas.

The nature (size, molecular weight, etc.) of particles ultimately formed depends upon conditions such as the residence time of the particles in the postflame gases, the temperatures to which these particles are subjected as they traverse this zone, and the degree to which

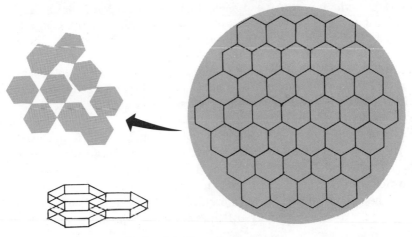

Figure 24. Growth of particles.

they are subjected to oxidizing or reducing conditions while still at
elevated temperatures. If the particles encounter oxidizing conditions,
they should oxidize if the temperatures are high enough:

$$C(s) + O_2 \rightarrow CO_2 \qquad\qquad (56)$$

whereas if the environment is reducing, gasification reactions can be
expected.

$$C(s) + CO_2 \rightarrow 2CO \qquad\qquad (57)$$
$$C(s) + H_2O \rightarrow CO + H_2 \qquad\qquad (58)$$

Note that for simplicity the particle is depicted as pure carbon in Reactions (56) through (58). In reality it is CH_y where $y \ll 1.0$.

The reaction of carbonaceous particles [C(s) denotes solid carbon]
with a gas is a heterogeneous or interfacial process.[38,39,40] Even if the
particles are uniformly distributed throughout the gas they are not
premixed in a molecular sense. Discrete clumps or aggregates of particulate matter are dispersed in a gas and the rate of reaction between
that particulate matter and components of the gas depends not only
on chemical factors but also upon the rate at which the two are brought
into intimate contact. Mechanistically the gaseous species must diffuse
from the bulk phase to the particle surface. Once there they may be
absorbed or react directly. After reaction the products must be desorbed before they may diffuse away from the surface to make room
for subsequent reaction of the particle with other gaseous molecules.
Thus a sequence of steps is involved: diffusion, absorption, reaction,
desorption and diffusion. The slowest of these steps controls the over-

all rate at which particles can react. Thus either oxidation or gasification of particles tends to be a relatively slow process. The reaction rate of an interfacial reaction is proportional to area. The length of time to completely oxidize or gasify a particle is related to initial particle mass. The larger the particle the more unfavorable the ratio of surface area to mass (all other things being equal) and hence the longer the time it must remain at elevated temperatures for complete reaction.

Several mechanisms for the production of particulate matter have been described above. Perhaps one of the most important classes of particles produced are the very small ones, that is, those with diameters 0.01–0.04 microns (refer to Figure B-5 in Appendix B). These small particles are termed combustion nuclei. Recent studies[41] have shown that they are produced by both premixed and diffusion flames. Furthermore, in the case of premixed flames they are emitted by lean as well as by rich flames.

Much is still unknown about the nature of these particles. For example, it is not known whether they form early in the combustion process and are partially consumed in the later stages of combustion, or whether their formation occurs in the postflame gases. Moisture does appear to greatly increase their concentration.[41] Also, their chemical composition is as yet unknown. Conceivably they could be mostly moisture-condensed on relatively stable flame ions that persist into the postflame gases. Alternatively, they could also result from reactions of acetylenic-type intermediates also present in the flame. In this latter case they would be composed largely of carbon and hydrogen. They are of particular interest because it is these small combustion nuclei that appear to coalesce or coagulate in the atmosphere to produce particles 0.1–0.5 microns in size. It is these latter particles that produce noticeable light scattering in the atmosphere.

2.9 CHEMICAL AND PHYSICAL PROCESSES IN THE EFFLUENT GASES

Thus far a variety of different chemical reactions that occur in the precombustion, flame, and postflame regions have been considered. The term *postflame* could be applied to all reactions that occur after the flame. It is desirable that a distinction be made between the recombination and pyrolytic reactions that have already been discussed in the postflame section and the myriad of other reactions that can subsequently occur in the effluent gases.

The possible reactions that may occur in the effluent include reactions between gaseous species, condensation of materials, liquid or solid phase reactions in the condensed material, and reaction of gaseous or condensed material with the walls of the effluent container. Before embarking on a discussion of these possibilities, it is worth considering their importance.

All of the reactions that occur in a combustion system are important insofar as they may alter the chemical and physical natures or the concentrations of trace species ultimately emitted to the atmosphere. Reactions in the effluent gases have the potential for doing just this. The nature of the reactions that do occur and their rates depend upon the conditions that prevail in the effluent, that is, upon the temperature, pressure, and residence time. These are variables over which the designer and to some extent the operator have control. Practical examples of effluent systems in which important chemical and physical processes do occur are smoke stacks of large stationary combustion processes and the muffler and exhaust pipes of an automobile.

The discussion in this section will be restricted to those reactions that occur simply because the flowing effluent gases are confined for the period of time during which they are conveyed to the point of discharge. The subject of effluent treatment processes and their associated reaction will be deferred until the next chapter.

Chemical Reactions of Gaseous Species

The nature of the various reactions occurring in the precombustion, flame and postflame regions have been discussed in the preceding sections of this chapter. It was seen that a species may be produced in one region that is consumed or otherwise modified in a subsequent region. The region or zone under current consideration is the final stage in the sequence of stages that constitute a combustion process. What is not reacted here will be emitted to the atmosphere, and whatever is produced by that which does react will also be emitted.

The effluent gases entering this zone contain a wide variety of species ranging from unreacted fuel to partial and complete reaction products. In some cases, reactions between different species are possible and occur so long as conditions are favorable.

It was noted that nitric oxide is the most stable oxide of nitrogen at the high temperatures encountered in combustion. On the other hand, at ambient temperature, nitrogen dioxide is the more stable oxide. Equilibrium considerations predict that when both oxygen and nitric oxide are present in the effluent gas, the nitric oxide will be oxidized to nitrogen dioxide as the gases cool. When these conditions prevail, this reaction does occur. However, the rate is low, and in most practical combustion devices the residence time is much too short for much conversion to occur. Generally, the concentration of NO_2 is well under 10% of the total NO_x on a mole basis.

A similar situation is encountered in the case of sulfur oxides. The dioxide is stable at flame temperatures, whereas the trioxide is more stable at ambient conditions. Again, the homogeneous gas phase reaction of SO_2 with oxygen is too slow to produce much SO_3. This reaction can be promoted by various materials that catalyze the reaction. This will be discussed later.

Reactions can occur between gaseous species. For example, sulfur dioxide can be reduced by carbon monoxide. It is reported that the principal product of this reaction is CO_2 with traces of carbonyl sulfide (COS) and carbon disulfide (CS_2) also being produced.[42]

Other reactions also can occur. For example, the water-gas shift reaction shown in Equation (47) can produce an adjustment in the relative levels of CO, H_2, H_2O and CO_2. The reaction will continue to occur so long as the temperature is sufficiently high. Once the level has dropped below a certain value, significant reaction ceases and the concentrations of the four species cited are essentially "frozen."

Condensation

Water is one of the major products of the combustion of fossil fuels. Figure 2 shows that depending on the equivalent ratio and the fuel composition anywhere from less than a half to more than two pounds of water may be produced per pound of fuel that is burned. At the high temperatures characteristic of combustion, this water remains in the vapor state.

Water is one of the condensable components present in the effluent gas. It is condensable insomuch as at STP conditions it is a liquid rather than a gas. Condensation will occur if the temperature of the effluent gas drops to or below the dew point; that is, the point at which the gas is saturated with water vapor. As can be seen from Figure 25, the dew point for the combustion of pure hydrocarbon fuels is relatively low. It ranges from about 100°F to a maximum of less than 140°F, depending upon the hydrogen-carbon ratio (y) of the fuel and the overall equivalence ratio. Normally, the temperature of the effluent gas would not drop to this level before it is discharged to the atmosphere. Even when it is discharged, condensation may not occur if the dew point is not reached due to mixing of the effluent gases with the relatively dry atmosphere.

The condensation of water vapor present in the effluent gases from combustion processes does occur under certain conditions and is a phenomenon undoubtedly familiar to most readers. It is particularly

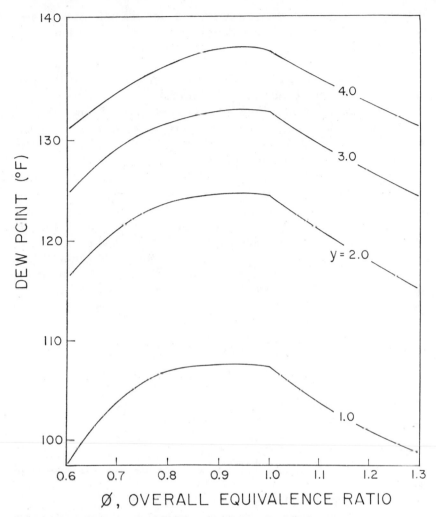

Figure 25. Relationship between dew point, overall equivalence ratio and H/C ratio of the fuel. Adapted from information in Reference 43. Equivalence ratio computed from air-fuel ratios.

likely to occur on cold days and appears as a white plume emanating from tailpipes of automobiles and stacks of stationary sources. This phenomenon should not be confused with the emission of particulate substances, which also produce an opaque plume. The color in this latter case usually ranges from a light gray to black.

It should be noted that the dew points shown in Figure 25 are computed from the concentrations of water produced by combustion alone.

Water may be present from other sources, for example, water vapor present in the intake air of the combustion process or water injected at some point in the combustion process to control nitrogen oxides. This "excess" or noncombustion water will raise the dew point somewhat; that is, as the effluent gases cool, condensation will occur at temperatures higher than those indicated in Figure 25.

The dew points shown in Figure 25 are based on ideal considerations. No account is taken of other species that may be present in the effluent gases of actual combustion processes. In a number of cases, these species may alter the actual dew point.

Particulate substances serve as condensation nuclei and thereby facilitate condensation. Particles are present in the effluent gases as discussed earlier. The source of some of these is the inlet air while others arise from pyrolysis reactions or from additives in the fuel. In general, the presence of particles will not alter the dew point measurably though as noted they do provide a site for condensation. When the particles are hydroscopic, that is when they absorb moisture at relative humidities less than 100%, they can remove water vapor from the effluent gases at temperatures higher than the dew point. Some salts that may arise from fuel additives exhibit this type of behavior.

Probably the best example of a hydroscopic substance that can alter the dew point is that of sulfur trioxide. It was noted earlier (Section 2.7) that most of the sulfur present in a fuel is oxidized to sulfur dioxide. Usually only a small fraction is converted to sulfur trioxide, the anhydride of sulfuric acid. The sulfur trioxide hydrolyzes with the water vapor to form sulfuric acid. The temperature where this mist first forms as the gases are cooled is called the acid dew point. Figure 26 shows the acid dew point as a function of the sulfur trioxide concentration in the effluent gases. Comparison of the ranges of dew point and acid dew point temperatures in Figures 25 and 26 illustrates the large increase resulting from the presence of relatively small amounts of sulfur trioxide. Figure 26 also shows the weight percentage of acid in the condensate and a measure of the corrosiveness of this condensate on steel.

Dilute acid is, of course, more corrosive than more concentrated acid. Thus, when significant quantities of sulfur trioxide are present, it is generally desirable to keep the temperature of the effluent gases high to minimize corrosion of structural surfaces from whatever acid mist is formed.

A related subject, which will be discussed in the next section and then again in Chapter 3, is the catalytic oxidation of sulfur dioxide. Some solid substances that may be by-products of combustion or that

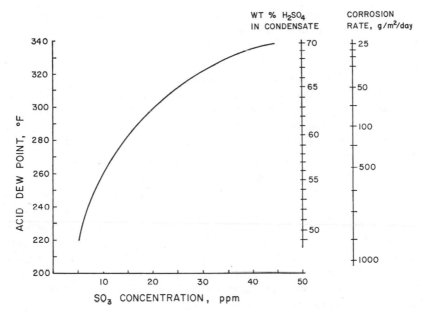

Figure 26. Acid dew point. Constructed from information in Reference 44.

may be intentionally placed in the effluent to promote the oxidation of other substances, also oxidize SO_2 to SO_3. When this occurs, the SO_3 concentration in the vapor phase increases as does the acid dew point. Another way of looking at this is that if the effluent gas conditions are such that when initially the operating point corresponding to temperature and SO_3 concentration falls above the convex line shown in Figure 26, no acid mist will be formed. If, now, the gas temperature is held constant while the vapor phase SO_3 concentration increases due to catalytic oxidation of SO_2, the operating point moves horizontally to the right until the acid dew point line is reached. At this condition, an acid mist forms. Further increases in the vapor phase SO_3 concentration will simply result in the formation of more and more concentrated acid at higher and higher temperatures. Spatially, this means that since the effluent gas is cooling, acid mist will be formed further and further upstream.

Thus far, two important phenomena have been identified that are associated with the condensation of moisture in the effluent gases. First, an opaque plume may result. Under most conditions, if water is the only constituent present, this is not detrimental to the environment. It may remind people of plumes that do contain particles and other pollutant species and thus be an object of criticism. There are some conditions in which the emission of water *per se* can be detrimental. One occurs when atmospheric conditions are such that icing is produced on power lines or highways downwind from a large combustion source. Another has been observed[45, 46] in arctic areas where the water of combustion freezes in the very cold air to form a fog of tiny ice crystals. When this does occur, a city or an airport is enveloped in a fog produced as a by-product of its own normal and necessary activities (space heating and the operation of mobile sources). When it is desirable to prevent condensation of moisture, heating (or reheating) of the effluent gases can be effective. When this is done, the hot moisture-laden gases can be diluted with a greater mass of drier atmospheric air before the temperature that would have produced condensation without as much dilution is reached.

The second phenomenon identified is corrosion that occurs when the condensate is acidic. Again, the solution is to maintain the temperature of the effluent gases at a high enough level to prevent condensation.

One additional process that may accompany condensation is absorption of soluble combustion products by the condensate. This is true whether the condensate is acidic or not. Whether the soluble species is in the gas or the liquid phase may or may not be of consequence environmentally. However, it does constitute a problem to the investigator who is attempting to measure the amounts of water-soluble combustion products such as the aldehydes and other carbonyls. If the gaseous sample is subject to cooling, condensate will form on the sample vessel walls and other surfaces such as filters. This condensate will absorb soluble species and reduce their gas phase concentration. Problems such as this explain why so little is currently known about some of the very important emission products of combustion systems.

A final note in this discussion of condensation. Water is not the only condensable substance in effluent gas. When certain fuel additives or other trace species are present in the fuel, metallic salts or oxides may be produced. These may be vapors at combustion temperatures. As the effluent gases cool, they condense or sublime, resulting in deposits on the vessel surfaces or, as the next chapter will reveal, on the surfaces of catalysts placed in the effluent stream. These deposits may be detrimental to the vessel (corrosive) and to any catalysts present (poisonous). Thus, control of effluent temperature may be an important method of preventing these detrimental effects.

Hydrocarbons and other species may also condense. Only the first few species in the paraffin series are gases at STP; the higher members are liquids and waxy solids. Other more complex hydrocarbons, such as the polynuclear aromatic discussed earlier, are also solids at STP.

As the effluent gases cool, these organic vapors condense on any surfaces present, including vessel walls and particles suspended in or entrained by the turbulent flowing effluent gases. Composite particles, which appear to consist of a nucleus of mineral or carbonaceous material with a somewhat transparent coating of a hydrocarbon substance, result.

Reactions with Surfaces

One type of reaction that can occur between effluent species and the surfaces of the effluent system has already been mentioned. This is the corrosive attack of acid mist. It was noted that the temperature at which acid mist begins to form and the acidic concentration of the mist both depend upon the vapor phase concentration of SO_3. Some materials that may be present on the surfaces of the effluent system may act as a catalyst for the oxidation of SO_2 to SO_3. One of these is vanadium pentoxide, V_2O_5. Vanadium pentoxide is a product of the combustion of vanadium-bearing fuels, for example, residual fuel oils. It may condense on the walls of the effluent system and promote the formation of an acid mist where none would otherwise form.

Another interfacial phenomenon of a quite different nature involves the deposition and agglomeration of particulate matter. Figure 27 shows the detailed morphological structure of a small particle when magnified 11,250x with a scanning electron microscope. This particle, which contains bromine and lead, is from the effluent of a spark-ignition engine. Its "fluffy" structure probably resulted from the gas phase collision of many smaller particulate nuclei. By contrast, the particle shown in Figure 28 is a much larger particle, probably resulting from the build-up of material on the walls of the effluent system. The finer structure reveals the impingement of many smaller particles upon the larger one. Large particles such as this can continue to grow on a surface until a momentary period of higher effluent gas velocity or temperature causes them to be dislodged and appear in the effluent. The result is a storage effect, that is, the rate of emission of the elements in the particle (mass/unit time) does not coincide with the rate of ingestion of these same elements in the fuel. They are ingested continuously with the fuel but their concentration appears as pulses or spikes in the effluent. The same can be said for the case of a molecular species, which is synthesized in the combustion process and stored in the effluent deposits. In either of these cases, a mass balance cannot be obtained over the entire combustion process unless the storage (accumulation) term is accounted for or averaging is done over a long enough period of time and a sufficient number of operating conditions.

Figure 27. Morphological structure of small particle, 11,250x magnification with scanning electron microscope. Courtesy, Dow Chemical Corp.

Figure 28 shows one example of the type of large particle that can be produced in the effluent of a combustion system. Acidic smut is another example. This is large particles that can form when acid mist deposits on and corrodes the metallic surfaces of the effluent system. A metallic sulfide is formed, which absorbs both carbonaceous material

Figure 28. Large particle, 5000x magnification. Courtesy, Dow Chemical Corporation.

and SO_2 and ultimately flakes off. The smut particles are quite large, and they settle on surfaces downwind from the combustion source, leaving a black smear when rubbed. Acid smut is most likely to form where the chimney or stack is uninsulated or otherwise poorly designed so that the effluent gas temperature drops to the point where acid mist forms. This is most likely to be a problem with liquid or solid fuels containing significant quantities of sulfur.

2.10 SUMMARY

The basic principles of combustion have been introduced in this chapter. The approach has been to follow the various reactions of the working fluid. Initially, this working fluid is an unreacted mixture of fuel and air. Subsequently, it undergoes a sequence of precombustion, combustion, postflame and effluent reactions. A variety of chemical species are formed, some of which undergo reactions in subsequent parts of the combustion process. Others, once formed, do not undergo further reaction and are subsequently exhausted to the atmosphere. The effluent gases contain a variety of chemical species, some gaseous, some liquid, some solid and some combinations of these, for example, a gas adsorbed on a particle.

Considerable attention has been devoted to the modification of the types and quantities of these trace species in the effluent by controlling parameters of the primary combustion process such as the air-fuel ratio, etc. At the same time, relatively little attention was devoted to the modification of the effluent gases by selection of and control over secondary combustion processes. This is the subject of the next chapter.

REFERENCES

1. Fristrom, R. M., and A. A. Westenberg. *Flame Structure* (New York: McGraw Hill Book Co., 1965).
2. Strehlow, R. A. *Fundamentals of Combustion* (Scranton, Pa.: International Textbook Co., 1968).
3. Williams, F. A. *Combustion Theory—The Fundamental Theory of Chemically Reacting Flow Systems* (Reading, Mass.: Addison Wesley Publishing Co., 1965).
4. Lewis, B., and G. von Elbe. *Combustion—Flames and Explosions* (New York: Academic Press, 1961).
5. Gaydon, A. G., and H. G. Wolfhard. *Flames—Their Structure, Radiation and Temperature,* 3rd ed. (London: Chapman Hall Ltd., 1970).
6. Knox, J. H. "A New Mechanism for the Low Temperature Oxidation of Hydrocarbons in the Gas Phase," *Combust. Flame,* 9, 297 (1965).
7. Minkoff, G. J., and C. F. H. Tipper. *Chemistry of Combustion Reactions* (London: Butterworths, 1962).
8. Tipper, C. F. H., Ed. *Oxidation and Combustion Reviews,* Vol. 1. (London: Elsevier Publishing Co., 1965).
9. Wesley, F. "A Supplementary Bibliography of Kinetic Data on Gas Phase Reactions of Nitrogen, Oxygen and Nitrogen Oxides," U.S. National Bureau of Standards Special Publication No. 371 (February 1973).
10. Wesley, F. "A Bibliography of Kinetic Data on Gas Phase Reactions of Nitrogen, Oxygen and Nitrogen Oxides," U.S. National Bureau of Standards (August 1971). Available from National Technical Information Service, Springfield, Va. 22151. No. NTIS COM-71-0084.

11. Baulch, D. L., D. D. Drysdale, D. G. Horn, and A. C. Lloyd. *Evaluation of Kinetic Data for High Temperature Reactions—Homogeneous Gas Phase Reactions for the H_2-O_2 System,* Vol. 1. (London: Butterworths, 1972). Vol. 2 was published in the fall of 1973.

12. Steffensen, R. J., J. L. Agnew, and R. A. Olsen. "Combustion of Hydrocarbons Property Tables," Engineering Bulletin of Purdue University, Engineering Extension Series No. 122, Lafayette, Indiana (May 1966).

13. Singer, J. M., E. B. Cook, M. E. Harris, V. R. Rowe, and J. Grumer. "Flame Characteristics Causing Air Pollution—Production of Oxides of Nitrogen and Carbon Monoxide," U.S. Bureau of Mines, Report of Investigation No. 6958 (1966).

14. Starkman, E. S., and H. K. Newhall. "Thermodynamic Properties of Methane and Air, and Propane and Air," presented at the SAE Mid-Year Meeting, Chicago, Ill., May 1967, Paper No. 670466.

15. Edwards, J. B. "Interfacial Phenomena in Combustion Systems," presented to the Dow Discussion Group on Interfacial Science, Midland, Mich. (November 1970).

16. Lee, K. B., M. W. Thring, and J. M. Beer. "On Rate of Combustion of Soot in a Laminar Flame," *Combust. Flame,* 6, 137 (1962).

17. Williams, J. R., A. S. Shency, and J. D. Clement. "Theoretical Calculations of Radiant Heat Transfer Properties of Particle-Seeded Gases," NASA CR-1505, (February 1970).

18. Ehigo, R., N. Nishiwaki, and M. Hirata. "A Study on the Radiation of Luminous Flames," *Proc. Eleventh Symp. (International) on Combustion,* Berkeley, Cal. (Pittsburgh, Pa.: Combustion Institute, 1967) p 381.

19. Liebert, C. H., and R. R. Hibbard. "Spectral Emittance of Soot," NASA TDN-5647 (February 1970).

20. Fenimore, C. P. "Formation of Nitric Oxide in Premixed Hydrocarbon Flames," *Proc. Thirteenth Symp. (International) on Combustion,* Salt Lake City, Utah (Pittsburgh, Pa.: Combustion Institute, 1971) p 373.

21. Newhall, H. K., and S. M. Shahed. "Kinetics of Nitric Oxide Formation in High Pressure Flames," *Proc. Thirteenth Symp. (International) on Combustion,* Salt Lake City, Utah (Pittsburgh, Pa.: Combustion Institute, 1971) p 381.

22. Blumberg, P., and J. T. Kummer. "Prediction of NO Formation in Spark-Ignited Engines—An Analysis of Control Methods," *Combust. Sci. Technol.* 4, 73 (1971).

23. Newhall, H. K. "Kinetics of Engine-Generated Nitrogen Oxides and Carbon Monoxide," *Proc. Twelfth Symp. (International) on Combustion,* Poitiers, France (Pittsburg, Pa.: Combustion Institute, 1969) p 603.

24. Westenberg, A. A. "Kinetics of NO and CO in Lean, Premixed Hydrocarbon-Air Flames," *Combust. Sci. Technol.* 4, 59 (1971).

25. Levy, A., E. L. Merryman, and W. T. Reid. "Mechanisms of Formation of Sulfur Oxides in Combustion," *Environ. Sci. Technol.* 4, 653 (1970).

26. Merryman, E. L., and A. Levy. "Sulfur Trioxide Flame Chemistry—H_2S and COS Flames," *Proc. Thirteenth Symp. (International) on Combustion,* Salt Lake City, Utah (Pittsburg, Pa.: Combustion Institute, 1971) p 427.

27. Singh, T., and R. F. Sawyer. "CO Reactions in the Afterflame Region of Ethylene/Oxygen and Ethane/Oxygen Flames," *Proc. Thirteenth Symp. (International) on Combustion,* Salt Lake City, Utah (Pittsburg, Pa.: Combustion Institute, 1971) p 403.

28. Smith, D. S., and E. S. Starkman. "A Spectroscopic Study of the Hydroxyl Radical in an Internal Combustion Engine," *Proc. Thirteenth Symp. (International) on Combustion,* Salt Lake City, Utah (Pittsburg, Pa.: Combustion Institute, 1971) p 439.

29. Tebbens, B. D., J. F. Thomas, and M. Mukai. "Particle Air Pollutants Resulting from Combustion," ASTM Symp. on Air Pollution Measurement Methods, Los Angeles, Cal. (Oct. 1962) ASTM Special Publication No. 352.

30. Ferguson, R. E. "An Isotopic Study of Carbon Formation in Hydrocarbon Flames," *Combust. Flame,* 1, 431 (1957).
31. Chakraborty, B. B., and R. Long. "The Formation of Soot and Polycyclic Aromatic Hydrocarbons in Diffusion Flames—Part 1," *Combust. Flame,* 12, 226 (1968).
32. Chakraborty, B. B., and R. Long. "The Formation of Soot and Aromatic Hydrocarbons in Diffusion Flames—Part III," *Combust. Flame,* 12, 469 (1968).
33. Commins, B. T. "Formation of Polycyclic Aromatic Hydrocarbons During Pyrolysis and Combustion of Hydrocarbons," *Atmos. Environ.,* 3, 565 (1969).
34. Davies, R. A., and D. B. Scully. "Carbon Formation from Aromatic Hydrocarbons," *Combust. Flame,* 10, 165 (1966).
35. Homann, K. H., and H. G. Wagner. "Chemistry of Carbon Formation in Flames," *Proc. Roy. Soc. (London),* A 307, 141 (1968).
36. Chakraborty, B. B., and R. Long. "The Formation of Soot and Polycyclic Aromatic Hydrocarbons in Diffusion Flames—Part II," *Combust. Flame,* 12, 238 (1968).
37. Homann, K. H. "Carbon Formation in Pre-Mixed Flames," *Combust. Flame,* 11, 265 (1967).
38. McLintock, I. S. "The Effect of Various Diluents in Soot Production in Laminar Ethylene Diffusion Flames," *Combust. Flame,* 12, 217 (1968).
39. Fenimore, C. P., and G. W. Jones. "Comparative Yields of Soot from Pre-Mixed Hydrocarbon Flames," *Combust. Flame,* 12, 196 (1968).
40. Lee, K. B., M. W. Thring, and J. M. Beer. "On Rate of Combustion of Soot in a Laminar Soot Flame," *Combust. Flame,* 6, 137 (1962).
41. Fissan, H. J., D. B. Kittelson, and K. T. Whitby. "Measurement of Aerosols Produced by a Propane-Air Flame in a Controlled Environment," report prepared by the Particle Technology Laboratory, Mechanical Engineering Department, University of Minnesota and submitted to the Air Pollution Control Office, U.S. Environmental Protection Agency, August 1972. Particle Laboratory Publication No. 190.
42. Ryason, P. R., and J. Harkins. "Studies on a New Method of Simultaneously Removing Sulfur Dioxide and Oxides of Nitrogen from Combustion Gases," *J. Air Poll. Control Assn.,* 17, 796 (1967).
43. D'Alleva, B. A. "Procedures and Charts for Estimating Exhaust Gas Quantities and Composition," General Motors Research Publication No. 372 (May 1960).
44. The data used in the construction of this figure was taken from the following sources: Corrosion data was obtained from N. D. Tomashov, *Theory of Corrosion and Protection of Metals* (New York: McMillan Co., 1966) p 510. The original reference for this corrosion data is G. V. Akimov. *Fundamental Studies on Corrosion and Protection of Metals* (Moscow, Russia: The State Scientific and Technical Publishing House for Literature on Ferrous and Non-Ferrous Metallurgy, 1946). Acid dew point data was obtained from *Steam—Its Generation and Use* (New York: The Babcock and Wilcox Co., 1955). The reader is also referred to the following note. F. H. Verhoff and J. T. Banchero. "A Note on the Equilibrium Partial Pressures of Vapors above Sulfuric Acid Solutions," *A.I.Ch.E. J.,* 18, 1265 (1972). These investigators point out that in the past acid dew points have been based on thermodynamic equilibrium calculations and may contain considerable error. They recommend that dew point predictions be based on experimental data.
45. Benson, C. S. "Ice Fog—Low Temperature Air Pollution," Cold Regions Research and Engineering Laboratory, Research Report 121. Corps of Engineers, U.S. Army, Hanover, New Hampshire (June 1970).
46. "Annual Report (1970–71) Geophysical Institute, University of Alaska, College, Alaska, 99701.

3

SECONDARY COMBUSTION PROCESSES

3.1 INTRODUCTION

The concepts of primary and secondary combustion zones were introduced in Section 2.6. Briefly, the primary combustion zone is where the major portion of the chemical potential energy stored in the fuel is released. This release takes place in a flame. Precombustion reactions precede this flame, and postflame reactions follow it. Secondary combustion processes, if they are present at all, occur downstream of the postflame zone as was shown in Figure 17.

When there are no constraints placed upon the types or quantities of chemical species that can be freely discharged, there is no particular incentive to treat the exhaust and thereby chemically modify its composition. A possible exception is the case of excessively rich combustion. When such constraints do exist, there are a variety of effluent gas treatment techniques that may be employed. Many of these techniques, such as baghouse filters, electrostatic precipitators, adsorbers and absorbers, have little in common with the subject at hand, combustion. Consequently, they will not be discussed here.

Three secondary treatment processes that are closely related to combustion and, hence, are referred to as secondary combustion processes will be considered in this chapter. These are thermal reactors, afterburners and catalytic reactors. These processes are usually used to achieve high overall combustion efficiencies as defined by Equation (34). They should not be confused with the various reactions discussed previously in Section 2.9 and which normally occur in the effluents of most combustion processes. Secondary combustion processes include homogeneous and heterogeneous, flame and nonflame reactions.

A thermal reactor is present in virtually all cases where secondary air is added to the postflame gases of a combustion process. Examples are the addition of secondary air to oxidize volatiles above a bed of burning coal, multiple chamber incineration, and injection of air into the exhaust manifold of an automobile engine. A possible exception in which the addition of secondary air does not produce secondary combustion occurs when massive amounts of dilution air are added to relatively cool effluent gases in order to decrease their opacity. This, of course, is not likely to produce any reaction at all.

Afterburners differ from thermal reactors in that secondary fuel is usually added along with secondary air. This, in turn, results in a flame that in turn promotes the oxidation of other species in the main feedstream, which is either passed through the flame itself or mixed with the hot effluent thereof. Afterburners find application in the treatment of effluents from combustion as well as noncombustion processes. Specific examples will be discussed later in this chapter as well as in Chapter 5.

The subject of catalytic reactors is one of current interest since large numbers of automobiles will be equipped with these devices in the very near future. Their use to treat the effluent of combustion processes is not really new. They have been installed on internal combustion engines used in closed spaces (mines, warehouses, etc.) for years now. Catalysis was mentioned briefly in the preceding chapter. The mechanism by which reactions of this type occur differ significantly from most other topics that have been discussed. Consequently, they are considered in detail in this chapter.

This chapter will be devoted to an analysis of the available types of secondary combustion processes, to their applicability for reducing the quantities of trace species present in different types of effluents, and to the ways in which they can be effectively used.

3.2 THERMAL REACTORS

The objective of promoting thermal oxidation reactions in the postflame effluent gases is to complete the oxidation of carbon monoxide and of any residual fuel and/or partially oxidized hydrocarbons. Two examples of thermal oxidation reactions have been considered previously. First were the precombustion reactions occurring at relatively low temperatures upstream of the flame; second were the much higher temperature oxidation reactions that occurred in the flame itself. The temperature range usually encountered in thermal oxidation of effluent gases is intermediate between these two extremes, and is typically greater than the 800–1200°F range encountered in the precombustion region and also very much lower than the 3000°F+ temperatures achieved in most flames.

In general the oxidation of carbon monoxide is more difficult than the oxidation of hydrocarbon species. This is due principally to the relatively slow carbon monoxide oxidation rate. Below about 1200°F the oxidation rates of carbon monoxide and hydrocarbons are not sufficiently rapid to significantly reduce the quantities of combustibles usually encountered in the effluent of practical combustion systems. The upper temperature limit encountered in applied secondary combustion processes is usually about 1600–1800°F. This arbitrary upper temperature limitation is imposed to prevent the severe materials problems that would be encountered if the materials of which the reactor is made were allowed to rise to higher temperatures. The walls are not cooled as are the walls of devices that come in contact with flames. Consequently no quench layer forms and oxidation efficiency is thereby increased. There is an additional reason why temperatures in excess of 1600–1800°F are not used. By the time this temperature range is achieved, the rates of the elementary oxidation reactions are usually great enough so that greater gains in oxidation efficiency are made by improved mixing of the effluent with the secondary combustion air than by increased reaction temperature.

Whether thermal oxidation occurs in the combustion or the postflame gases it is always an exothermic process. The temperature rise that takes place if this oxidation occurs adiabatically is shown in Figure 29. Maintenance of the thermal oxidation within the desired temperature range requires a balance between the initial temperature of the postflame gas, the thermal energy liberated by oxidation of combustibles in this gas, the actual heat loss from the reactor and any cooling that takes place when lower temperature diluents or excess air are added. Since thermal oxidation in the effluent gases usually occurs at conditions approximating constant pressure, an enthalpy balance analogous to Equation (22) and allowing for any finite heat losses can be constructed to describe the process.

Thermal oxidation differs in yet another aspect from the normal combustion process. In the latter case certain constraints must be satisfied before the reaction can proceed to completion. In particular the ratio of fuel or oxidant to diluents must be such that the mixture composition lies within the flammable envelope shown previously in Figure 19. By contrast thermal oxidation reactions in the postflame effluent gases can and often do occur at combustible concentrations much below the lean flammability limit and also when the concentration of oxygen is much too low to sustain a flame. Rather than a conventional flame, thermal oxidation may proceed via homogeneous gas phase reactions. What is required in this case is that the effluent gases in the thermal reactor be maintained at reaction temperature (1200–

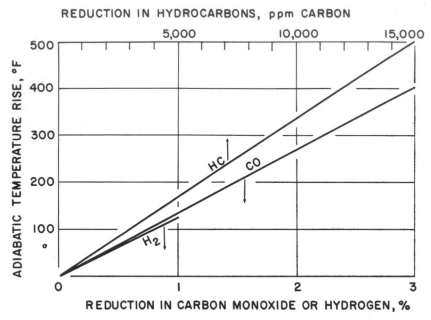

Figure 29. Adiabatic temperature rise accompanying oxidation of combustibles.

1600°F) for a time sufficiently long to allow the desired degree of conversion (oxidation) to occur.

There are basically two different types of thermal reactors used in conjunction with combustion systems. First is the type of reactor that is used downstream from a flame when the mixture is lean. In such cases the postflame gases usually contain relatively low concentrations of carbon monoxide and hydrocarbons as well as a low concentration of oxygen. Generally this oxygen is uniformly distributed throughout the effluent gases. When the initial gas temperatures are at least as high as the desired reactor temperature range (1200–1600°F), thermal oxidation can be promoted by carefully designing the effluent system to minimize heat losses and to allow sufficient residence time for the combustibles to oxidize at elevated temperatures. Sometimes secondary air is added as shown earlier in Figure 17. However, even when attempts are made to preheat the secondary air, its temperature is usually considerably less than the postflame gas temperature and its addition lowers the overall mixture temperature. Thus, the addition of secondary air to increase the oxygen concentration is usually not done for the case when the primary combustion stoichiometry is lean unless the air temperature is equal to or greater than the desired reaction temperature range. From a kinetic point of view, because the rates of oxidation are more sensitive to temperature than to oxygen concentration, the overall conversion is likely to decrease if the temperature drops appreciably in spite of an increase in oxygen concentration.

The second case of interest is the utilization of a thermal reactor to oxidize the relatively large quantities of combustibles produced when the primary combustion is rich. In this case the addition of secondary air is a stoichiometric necessity. Sometimes more than the minimum amount required to oxidize the combustibles is added because this excess will limit the overall temperature rise, which can be large when high concentrations of combustibles are present. The processes associated with the addition of secondary air as shown in Figure 17 are expanded upon in Figure 30. For the present, only the process entitled "Thermal Reactor" will be of concern. Note that an allowance must be made for mixing to occur between the secondary air and the high temperature postflame gases. This step is followed by thermal oxidation or pyrolysis reactions in the reactor. Pyrolytic reactions are likely to occur when the amount of secondary air added is insufficient to completely oxidize the combustibles or when the mixing is incomplete and not all of the postflame gases have an opportunity to combine with the available oxygen.

Mixing processes within thermal reactors are important. One of these processes, the mixing of secondary air with the postflame gases, has already been mentioned. There is still another type of mixing that must be considered. This is the degree to which mixing occurs between the various parcels of gas that are contained simultaneously within and in the process of flowing through the reactor. The case in which there is no mixing between successive elements of reacting fluid as they pass through the reactor is characterized as a plug flow reactor. The antithetical case, that is where mixing is perfect and therefore there can be no concentration gradients within the reactor, is called a well-stirred reactor.[1]

Plug flow is illustrated in Figure 31 and is sometimes referred to as slug flow or nonbackmix flow. Reactants enter at the left and flow through the reactor in an orderly fashion. Mixing can occur in the radial but not in the axial direction. Flow through the reactor can be divided into a series of elements or plugs, one of which is shown in Figure 31. The residence time τ for each of these elements is identical as it passes through the reactor. Residence time has the units of time and is defined as

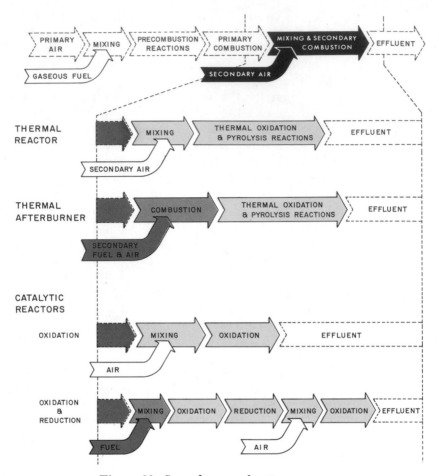

Figure 30. Secondary combustion processes.

$$\tau \text{ (time)} = \frac{V \text{ (volume)}}{V_i\left(\dfrac{\text{volume}}{\text{unit time}}\right)} \qquad (59)$$

where

   V = reactor volume
   V_i = volumetric feed rate

The reciprocal of residence time is space velocity, $S = 1/\tau$, and it, of course, has reciprocal time units. If the concentration of some combustibles, C, entering the reactor is $[C]_i$ and that leaving the reactor is $[C]_o$, the rate of oxidation can be generalized in terms of a global kinetic expression as follows.

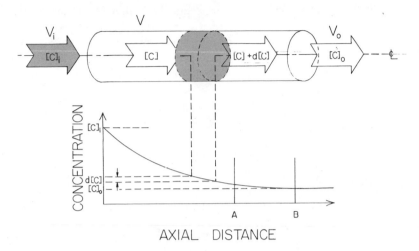

Figure 31. Plug-flow reactor.

$$r_o = -\frac{d[C]}{dt} = k_f\{[C]\,[O_2]\} \tag{60}$$

where

r_c — rate of oxidation of combustible C

$[O_2]$ = concentration of oxygen

k_f = rate constant for the oxidation reaction

$f\{\ \}$ denotes some functional relationship that depends on the nature of the combustible species, etc.

The negative sign indicates that the concentration of the combustible is decreasing. It is apparent from this analysis and from the concentration profile shown in Figure 31 that the degree of conversion depends on providing adequate reactor volume. If the reaction is terminated at point A rather than at point B, the degree of conversion will be less than if it is allowed to continue for the longer period of time. In dealing with secondary combustion devices it is common to define efficiency (conversion efficiency) in terms of the degree to which some component is oxidized or otherwise converted.

$$\eta_{conv} = \frac{[C]_i - [C]_o}{[C]_i} \times 100 \tag{61}$$

Note that this is a basically different method of defining efficiency than those adopted earlier for combustion and thermal efficiencies [refer to Equations (33) and (34), respectively].

A well-stirred or backmix reactor is one in which the reactants are instantaneously diluted as they enter the reactor. There are no concentration gradients within the reactor and the composition throughout the reactor is assumed to be identical to the composition of the reactor outlet stream. This is illustrated schematically in Figure 32. The

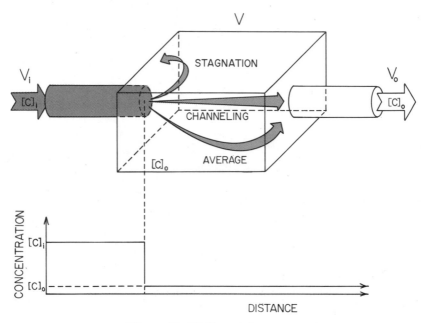

Figure 32. Well-stirred reactor.

residence time for material within the reactor is variable. Some parcels channel directly from the inlet to the outlet and thereby have a very short residence time, while others stagnate in turbulent eddies and have an abnormally long residence time as shown. As such, residence time does not have too much meaning for this case. Sometimes a mean or average residence time is determined.

$$\tau = \frac{V \text{ volume}}{V_o \left(\dfrac{\text{volume}}{\text{unit time}} \right)} \tag{62}$$

where
$\quad V =$ reactor volume
$\quad V_o =$ outlet volumetric flow rate

The outlet flow rate is used here because the exit stream represents conditions within the whole reactor.

Reaction of the combustibles in a well-stirred reactor occurs at the composition level of the outlet stream rather than over a series of continuously variable intermediate concentrations as was the case for plug flow. Thus the rate equation for a well-stirred reactor is:

$$r_c = -\frac{d[C]}{dt} = k_f\{[C]_o[O_2]_o\} \qquad (63)$$

The conversion efficiencies for the two cases are identical and are shown in Equation (61).

A number of important things are worth noting about plug flow and batch reactors. First, the fractional conversion

$$F_c = \frac{[C]_i - [C]_o}{[C]_i} \qquad (64)$$

achieved by one approach is not necessarily better than that achieved by another. The relative conversions depend upon the specific functional relationships in Equations (60) and (63) and upon other factors such as residence time, etc. The fractional conversion of one type reactor may be greater than, equal to or less than the other.

An additional difference between plug and well-stirred reactors can be illustrated by observing differences in the energetics of the reactions. In the case of an adiabatic plug flow reactor the temperature increases in monatonic fashion as fractional conversion increases through the reactor. This is, of course, not the case with a plug-flow reactor where by definition the temperature within the reactor T_r is uniform and equal to the outlet temperature T_o. The fraction of the energy released, F_e, is proportional to the fractional conversion of the combustibles and the heat of combustion, ΔH_n, of the combustibles. If the heat capacity of the reacting mixture is assumed constant over the temperature range involved then a straight line relationship must exist between the fractional energy release and the reactor temperature T_r (or T_o) as shown in Figure 33. When no reaction occurs, $F_e = 0$, and when all of the available combustible has reacted, $F_e = 1.0$.

An analogous situation exists for the fractional conversion of combustible F_c. In this case the shape of the curve depends upon the form of the kinetic expression for conversion of the combustibles. For a well-stirred reactor the global kinetic expression is typically of the form $Ae^{-E/RT_r}[CO]_o[O_2]_o$. At low conversions the curve is asymptotic to $F_c = 0$ due to the term $1/e \ (E/RT_r)$ which is very small at low temperature. At high conversions the curve becomes asymptotic to $F_c = 1.0$ because $[CO]_o \rightarrow 0$ and perhaps $[O_2]_o \rightarrow 0$ also. The result is a curve with an inflection as shown in Figure 34.

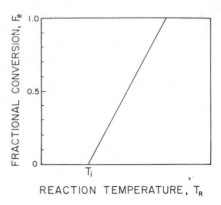

Figure 33. Relationship between fractional energy release and reactor temperature.

It is apparent that at any real operating point of the reactor the following relationship must be satisfied: $F_c = F_e$; or, restated, for adiabatic operation of the reactor the conversion of reactants must liberate precisely enough energy to produce the indicated temperature rise in the reactor. This may be illustrated graphically by combining the curves in Figure 33 and Figure 34. The result is shown in Figure 35.

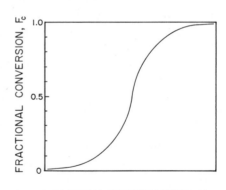

Figure 34. Relationship between fractional conversion of reactants in a well-stirred reactor and reactor temperature.

The kinetic and energy balance curves intersect at three points—A, B and C. It can be shown[2] that A and B are both stable operating points of the thermal reactor and correspond to low-temperature–low-conversion and high-temperature–high-conversion, respectively. C is an unstable point and the reactor will revert to either A or B. If the initial temperature, T_i', is sufficiently low, there will only be a single low-conversion operating point, D, as shown. Conversely, if the initial temperature is sufficiently high, there will also be only a single high-conversion operating point, point E in this case. The objective of em-

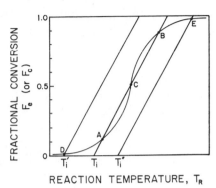

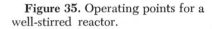

Figure 35. Operating points for a well-stirred reactor.

ploying a thermal reactor is to achieve high conversions such as those corresponding to points B and E. This is no great problem so long as the inlet temperature of a well-mixed reactor is high enough. However in practical situations, and particularly when transient inlet conditions are encountered, it will not always be possible to attain a sufficiently high T_i. Success in achieving consistently high conversions requires careful attention to reactor design and operating characteristics.

In concluding this discussion of thermal reactors it should be noted that practical reactors are neither truly plug flow nor well-stirred. They are generally somewhere between these two and can be designed to be more nearly like one than the other in order to achieve whichever operating characteristics are desired.

3.3 AFTERBURNERS

Thermal afterburners can be used to oxidize hydrocarbons, carbon monoxide and particulate matter present in the postflame gases of a combustion process. The major processes of an afterburner are shown in Figure 30. Note that there are both similarities and dissimilarities with the thermal reactor approach discussed previously. The thermal afterburner requires the addition of secondary fuel but the thermal reactor functions on the energy content present in the influent combustibles. Both approaches oxidize gases and particles by providing an environment wherein high temperature *and* adequate oxygen promote oxidation. Pyrolysis may occur if the former but not the latter of these two objectives is achieved.

An afterburner can be envisioned as a device in which the postflame gases are passed through a flame. This is noted as the combustion zone in Figure 30. The flame is created by the addition of secondary fuel to supplement the heating value of combustibles already present in the postflame gases. Secondary air may have to be added to create a flammable mixture. The energy released in the flame raises the gas temperature to the desired level and the addition of excess air with the fuel provides the desired oxygen concentration in the thermal oxidation zone.

The nature of the reactions that occur in the thermal oxidation and pyrolysis zone are very similar to those previously discussed in the thermal reactor section and will not be repeated here. If it is necessary to oxidize particulate matter, the afterburner must be designed with this in mind. It was pointed out earlier that particles are good radiators of thermal energy. A good radiator of energy is also a good absorber. Therefore, when particles must be oxidized, the basic afterburner design should promote the transfer of radiant energy to these particles. This may be accomplished by creating a flame that is luminous and/or by constructing the afterburner with refractory walls that can withstand sustained high temperatures and thereby radiate energy to the particulate matter contained in the reacting gases.

A schematic of a thermal afterburner is presented in Figure 36. The choice of secondary fuel depends, among other things, upon what is

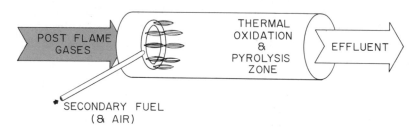

Figure 36. Afterburner.

readily available and what can be handled conveniently. Most commonly the choice is natural gas. The secondary burner may take many forms, and is shown here as a ring with multiple ports. If the secondary air is premixed with the fuel, the flame at each port is not unlike the Bunsen burner flame shown in Figure 4. If it is not premixed, then the fuel will burn in a diffusion flame, which will be discussed in greater detail later on. For the moment refer to Figure 57 for an illustration of a diffusion flame.

The temperature that must be sustained in the oxidation zone to completely oxidize the combustibles, and the necessary residence time, are interrelated. The higher the gas temperature the shorter the residence time required, and vice versa. The nature of the combustible species is also a factor. Some complex organic molecules crack readily as they pass through the flame zone and the fragments are easily oxidized while other species such as methane are much more refractory and require considerably higher temperatures for oxidation. Generally gas temperatures in the range of 1400–1600°F are desirable to oxidize commonly encountered organics. Another rule of thumb is to allow a 0.3 second residence time in the oxidation zone. This is only a guideline. The presence of large particles will obviously require longer residence times than when the particles are small or absent.

The discussion here has been focused upon the use of thermal afterburners to complete the oxidation of species in the effluent of combustion systems. There are many applications of afterburners other than treatment of effluents of combustion processes. They can be employed to oxidize organic species, both gaseous and particulate, present in the effluent of a variety of noncombustion processes, for example, the effluents from paint spray booths or from casting operations in foundries. The object here is not to list or to discuss these other applications in detail for these will be addressed in Chapter 5 but to explore those conditions that made the use of an afterburner an attractive method of effluent treatment.

An afterburner is attractive for application when the inlet temperature of the gas to be treated is too low for oxidation of the organic species present to occur, and when the concentration of these organics is so low that it exceeds the lean flammability limit for the particular mixture. Under these conditions the addition of fuel and its subsequent combustion can be used to raise the mixture to the temperature necessary for oxidation of all organic species present. It should be pointed out that even in the case where the mixture is combustible the addition of fuel in an afterburner may be desirable because it increases the final reaction temperature. However, when the influent to the afterburner is combustible the use of some type of flame arrestor on the afterburner inlet is essential to prevent the flame from propagating upstream.

The quantity of additional fuel that must be added depends upon the inlet temperature of the gas to be treated, on the mass of this gas and on its composition as well as on the desired reaction temperature. This additional fuel represents an economic penalty and it is desirable to minimize the quantity of secondary fuel which must be added. Arrangements whereby the inlet gases are heated by energy exchange with the hot effluent gases as shown in Figure 37 are often desirable and minimize the secondary fuel requirements. The extent to which heat exchange is desirable depends upon the relative costs of fuel *vs.* heat exchange surface.

The use of thermal afterburners to oxidize organic species in the effluent of combustion processes is rather straightforward. The afterburner is simply completing a process that was begun initially in the primary combustion stage. When afterburners are applied to effluents from noncombustion processes, care must be exercised in their applica-

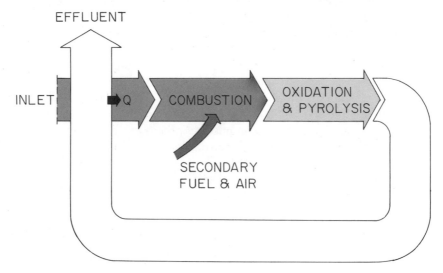

Figure 37. Energy conservation with an afterburner.

tion to insure that the products of oxidation or pyrolysis are not even less desirable than the initial species. This applies to effluents containing large quantities of sulfur-bearing compounds that will invariably be oxidized to SO_2. The same is true of halogenated species such as the freons. When these molecules are oxidized, the reaction products may include hydrogen chloride, hydrogen fluoride and phosgene, to mention only a few of the possibilities.

It is worthwhile to note that both afterburners and thermal reactors oxidize species via a sequence of intermediates. If the temperature and the residence time are insufficient for complete oxidation, partial oxidation products such as carbonyls may result.[3] This is another case where the products are less attractive than the reactants from the point of view of air pollution.

An additional factor that should not be overlooked with regard to secondary combustion processes such as thermal reactors and afterburners is that energy is released at a temperature lower than that occurring in the primary zone. In many cases this energy is not recoverable, that is it simply is lost as hot effluent gas without any useful work being performed. Even if it is possible to convert a portion of this energy into work, it is released at a lower temperature than would have been the case if it had been released during the primary combustion processes. The value T_c in Equation (33) is lower for secondary combustion than it is for most primary combustion processes. Therefore, the ultimate thermal efficiency (Carnot cycle efficiency) is lower for secondary than for primary combustion. In summary, secondary combustion can be used to increase combustion efficiency as defined by Equation (34) at the expense of thermal efficiency.

3.4 CATALYTIC REACTORS

A catalyst is a substance that promotes the rate of a chemical reaction without undergoing any permanent chemical change itself. Insofar as secondary combustion processes are concerned there are two cases of interest. The first is the use of oxidation catalysts to promote the complete oxidation of combustibles present in the effluent of the primary combustion process.

$$CH_4 + 2O_2 \rightarrow CO_2 + 2H_2O \tag{65}$$
$$2CO + O_2 \rightarrow 2CO_2 \tag{66}$$
$$2H_2 + O_2 \rightarrow 2H_2O \tag{67}$$

Another important type of reaction that may be promoted by the use of catalysts is the reduction of nitrogen oxide.

$$2NO + 2CO \rightarrow N_2 + 2CO_2 \tag{68}$$
$$2NO + 2H_2 \rightarrow N_2 + 2H_2O \tag{69}$$
$$4NO + CH_4 \rightarrow 2N_2 + CO_2 + 2H_2O \tag{70}$$

In Reactions (68) through (70) the CO, H_2 and the hydrocarbon, CH_4, are the reductants and the NO is reduced. Note that these reactions are different from the decomposition of NO.

$$2NO \rightarrow N_2 + O_2 \tag{71}$$

Reaction (71) is much more difficult to catalyze than (68–70) and requires much higher temperatures and longer residence times than are generally practical in treating combustion system effluents.

There are a number of other combustion effluent treatment processes that involve catalysts, for example, the catalytic removal of SO_2 from the effluent gases produced by combustion of high sulfur containing liquid and solid fuels. These processes are essentially complex chemical

processes to convert the sulfur oxides from their dispersed gaseous state
to a form more convenient for disposal or sale. This type of catalytic
treatment process is unlike the secondary combustion processes being
discussed here with respect to operating principles and conditions,
temperature, etc. Consequently catalytic removal of SO_2 from combustion effluent gases will not be discussed here. It is, however, discussed in some of the references in the Stationary Boiler section of
Chapter 5.

Theory

A catalyst was described earlier as a substance that promotes a
chemical reaction. In other words the catalyst increases the rate at
which a thermodynamically feasible reaction approaches equilibrium.
For example, consider Reaction (68).

$$2NO + 2CO \underset{k_b}{\overset{k_f}{\rightleftarrows}} N_2 + 2CO_2$$

k_f and k_b are the forward and reverse reaction rate velocities. A catalyst
for this reaction increases the magnitudes of both k_f and k_b and thereby
increases the rate at which the four species undergo mutual reaction
in either direction until equilibrium is achieved as defined by the
relationship

$$K = \frac{k_f}{k_b} = \frac{[N_2][CO_2]}{[NO][CO]} \tag{72}$$

Stated in another way, the presence of a catalyst lowers the temperature
required to achieve a given degree of conversion within a fixed reactor
volume and available time period.

A catalyst increases the forward and reverse reaction rates by lowering the activation energy separating reactants and products. This is
illustrated in Figure 38. The relationship of activation energy E_A for
the homogeneous reaction and the heat of reaction ΔH_R to the energy
associated with reactants and products is shown. For the reactants to
undergo a homogeneous (gas phase) reaction and be converted into
products they must pass along the path denoted by the solid line. The
high energy barrier that is encountered initially along this path results
in a relatively slow rate of reaction. If a catalyst for the reaction is
present, this energy barrier is reduced as shown. The catalyst provides
a new path denoted by the dotted line by promoting the formation of
a necessary intermediate, A^*, on or at its surface, that is, the interface
between the reactive gas and the solid catalyst. In essence the potential
energy barrier between reactants and products is lowered and the
reaction proceeds at a much faster rate.

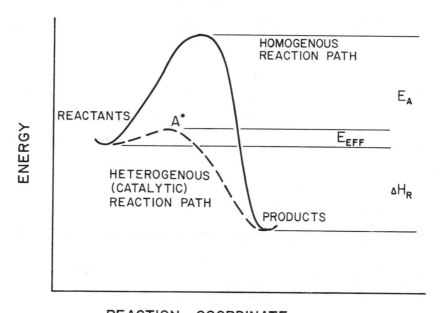

Figure 38. Reaction coordinate diagram.

Catalytic reactions proceed via a series of discrete steps or processes. Briefly these are:

1. diffusion of reactants from the gas phase to the vicinity of the catalyst surface
2. adsorption of one or more of the reactants on the catalyst surface
3. reaction of adsorbed species with other adsorbed species or with nearby gas phase molecules
4. desorption of the reaction products
5. diffusion of these product molecules from the reaction surface through the fluid boundary layer and into the bulk gas phase.

Each of these steps will be discussed in greater detail below.

The first step is diffusion of reactants from the bulk gas stream to the catalyst surface. Regardless of how fast the gas is flowing past the catalyst surface, the viscous nature of the gas will result in a boundary layer of finite thickness at the catalyst gas interface. The thickness of this boundary layer will depend upon the bulk gas velocity and the physical properties of the gas at the temperatures and pressures prevailing. Within this viscous boundary layer mass transfer occurs by molecular rather than turbulent diffusion. As will be apparent from later discussions molecular diffusion is also the predominant transport mechanism in the crevices and pores within the catalyst. Molecular diffusion is a relatively slow process compared to some other processes, such as the rates of many of the chemical reactions under consideration. Therefore it may be the overall rate controlling step in determining the overall conversion (oxidation or reduction) that can be achieved with a catalyst.

The rate of diffusion is described by Ficks' first law:

$$J = -D\frac{\partial C}{\partial X} \tag{73}$$

where J is the diffusive flux, D the diffusion coefficient and $(\partial C/\partial X)$ the concentration gradient. D is dependent upon temperature, pressure and the chemical composition of the gas phase as well as the species that is diffusing. The magnitude of the concentration gradient depends upon two major factors, first the thickness, X, of the viscous boundary layer and second the difference in concentration across this boundary layer. The thickness of the boundary layer is determined by hydrodynamic factors and generally decreases as the velocity of the bulk fluid flowing over the catalyst surface increases. The difference in concentration across the boundary layer depends upon:

1. the concentration of the diffusing molecular species in the bulk gas, and
2. the rate of adsorption or reaction of this species with the catalyst surface or species adsorbed thereon.

With this brief discussion of diffusion as background two things should be apparent. For a material to be catalytic one or more of the reactants must have an affinity for its surface. Second it is desirable to arrange the surface with respect to the flowing bulk gases so that the diffusive flux will be as great as is practically possible.

The second step in the overall catalytic mechanism is adsorption. As pointed out earlier one or more of the reactants in a reaction must be adsorbed by a surface for catalysis to occur. When a species is adsorbed it is in effect concentrated at that surface. This concept is illustrated in Figure 39. The surface is termed the adsorbent and the adsorbed species is the adsorbate. Due to differences in electronic structure of the adsorbate and of different gas molecules not all species are adsorbed equally. This is the situation in the case illustrated. If the plain species have any propensity to be adsorbed on the surface they are effectively blocked by the shaded ones, which have a much greater affinity for the particular surface. The greater local concentration of shaded species near the surface is readily apparent.

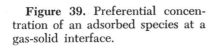

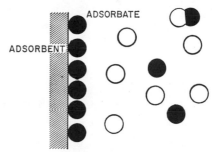

Figure 39. Preferential concentration of an adsorbed species at a gas-solid interface.

In addition to the consideration of whether or not a species is adsorbed or not, its orientation on the surface may not be uniform, particularly when it is not symmetric geometrically and it has an electrical dipole. This is illustrated in Figure 40. The orientation of the species, *i.e.*, which "end is up" or which down (contacting the surface), may well govern whether any reaction occurs and if one does the nature of that reaction.

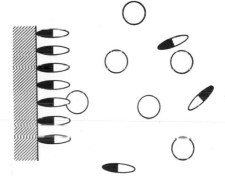

Figure 40. Preferential orientation of an adsorbed species at a gas-solid interface.

The adsorption of a molecule on a surface may be physical or chemical. In the case of physical adsorption the forces that bind a molecule to a surface include permanent dipoles, induced dipoles, quadrapole attraction and the well-known VanderWall's forces. The heat of physical adsorption is small (typically 2–6 kcal/mol) and there is no activation energy associated with this process. These factors are illustrated by the energy curve shown for physical adsorption in Figure 41. Several layers of adsorbed molecules may be built up, thus concentrating the adsorbate.

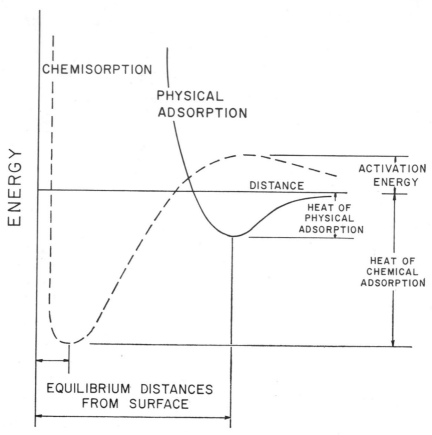

Figure 41. Energetics of chemical and physical adsorption at a gas-solid interface.

Insofar as catalysis is concerned chemisorption is a more important process than physical adsorption. When chemisorption occurs a rearrangement of elections occurs producing actual chemical bonds between the adsorbent and the adsorbate. Chemisorption is therefore limited to a single layer of adsorbate molecules, which is held much closer to the surface. Large amounts of energy (from twenty to a few hundred kcal/mol) may be released during chemisorption.

The chemisorption process may be considered as reversible or irreversible. Figure 42 illustrates both of these conditions. A reversible process is shown in the upper portion of the figure. Carbon monoxide is reversibly adsorbed by a metal oxide surface. The dotted line denotes a chemical bond. When the heat of adsorption, Δ, is subsequently returned, the carbon monoxide is desorbed with no net change to either

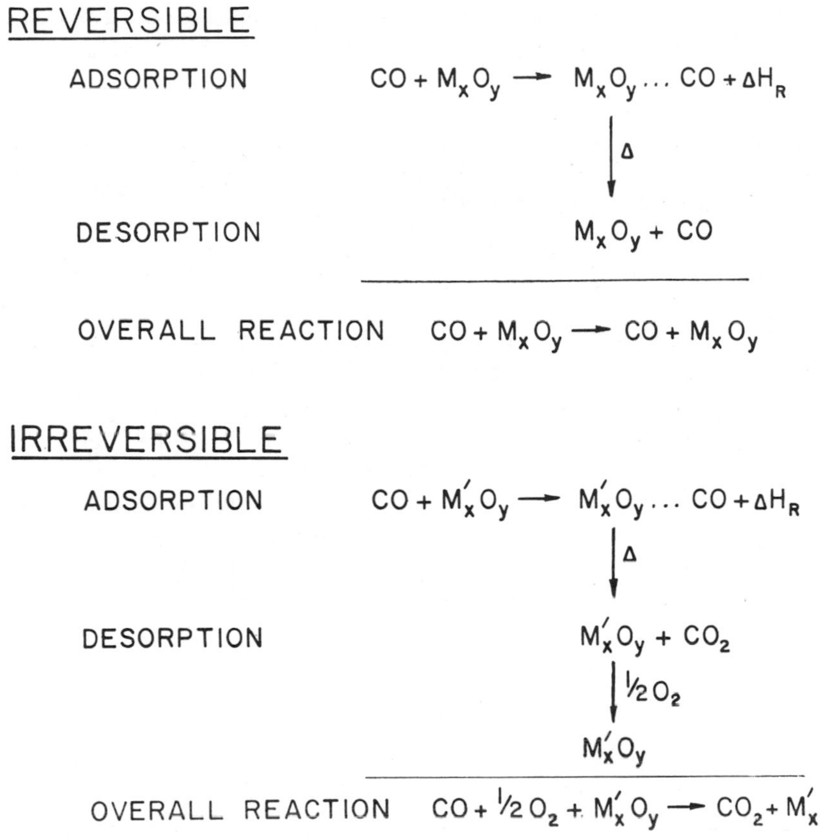

REVERSIBLE

ADSORPTION $CO + M_xO_y \longrightarrow M_xO_y \ldots CO + \Delta H_R$

$$\downarrow \Delta$$

DESORPTION $M_xO_y + CO$

OVERALL REACTION $CO + M_xO_y \longrightarrow CO + M_xO_y$

IRREVERSIBLE

ADSORPTION $CO + M'_xO_y \longrightarrow M'_xO_y \ldots CO + \Delta H_R$

$$\downarrow \Delta$$

DESORPTION $M'_xO_y + CO_2$

$$\downarrow \tfrac{1}{2}O_2$$

$$M'_xO_y$$

OVERALL REACTION $CO + \tfrac{1}{2}O_2 + M'_xO_y \longrightarrow CO_2 + M'_x$

Figure 42. Reversible and irreversible adsorption of carbon monoxide by a metallic oxide surface.

adsorbate or adsorbent. By way of contrast, if CO is irreversibly adsorbed by another metallic oxide as shown in the lower portion of Figure 42, application of heat results in desorption of carbon dioxide rather than carbon monoxide. An oxygen atom is transferred from the metallic oxide lattice to the adsorbate. Subsequent replacement of this oxygen atom by the addition of oxygen from a molecule of oxygen that has also diffused to the surface results in no net change to the adsorbent. This type of behavior is typical of those compounds called nonstoichiometric oxides in which the metal can assume a multiplicity of oxidation states. In this system such a catalyst may be said to promote a reaction without undergoing any net change. The catalyst shown in the lower portion of Figure 42 promotes the oxidation of carbon monoxide to carbon dioxide. In doing so it enters into the mechanism of the reaction. So long as there is a source to replace the oxygen lost by the lattice the catalyst undergoes no *net* change. If for some reason no source of oxygen is available, the metal oxide will cease to act as a catalyst when the available lattice oxygen at its surface is also depleted. Thus a surface is a catalyst only under specific conditions. In this case the specific conditions necessary for oxidation of CO to CO_2 are the presence of both CO and O_2.

The third step in the overall catalytic mechanism is the chemical reaction on the surface itself. The catalyst functions by providing an alternate path for the reaction with a lower potential energy barrier as was shown in Figure 41. The intermediate denoted as A^* was formed along the alternate reaction path. This intermediate may be quite different from the intermediate(s) formed along reaction paths corresponding to high temperature combustion. For example since the catalytic reaction occurs at a lower temperature, the reaction intermediates along this path may not even be thermally stable at the conditions required for flame reaction of the same species. Therefore, since the reaction mechanism is different and the intermediate species are different the product distribution obtained from the catalytic reaction of a given set of reactants may differ from the product distribution, which is obtained when these same reactants undergo flame reaction at much higher temperatures. Furthermore since other steps in the overall catalytic reaction mechanism, such as adsorption and desorption, are temperature-dependent the product distribution obtained from the catalytic reaction of a given set of reactants may vary with temperature. Often, in the case of catalysis, the choice of a catalyst and of operating conditions is governed as much by considerations of the desired product distribution as it is by the desire to achieve the greatest absolute magnitude of the reaction rate.

Reactions at a catalytic surface may be of various types. Two chemisorbed species may react with one another or one chemisorbed species may react with a nearby gaseous molecule, for example, with a molecule that is physically adsorbed in the layers above the chemisorbed species. The rate of these reactions is sometimes expressed in a modified Arrhenius form. The conventional Arrhenius equation $r = A\exp(-E/RT)$ for homogeneous reactions expresses the rate as a function of a preexponential term A, which is a measure of the number of effective collisions of gas molecules and an activation energy E. This equation can be modified as follows to express the rate of reaction on a catalyst surface: $r = A^l\exp(-E_{EFF}/RT)$. The modified preexponential term A^l is a measure of the number of active sites on the catalyst. The number of sites increases as the surface area available for adsorption increases. It may also be thought of as reflecting the increased local concentration of reactant molecules near the surface. The exponent is also modified. The effective activation energy, $E_{EFF} = (E - \Delta H_q)$, reflects the fact that when the reactants are chemisorbed by the catalyst the activation energy for the reaction is decreased by the heat of chemisorption, ΔH_q.

The final two steps in the overall catalytic mechanism are desorption and diffusion of product molecules from the catalytic surface into the bulk gas. The product molecule(s) must not have so high an affinity for the surface that they are tenaciously held and block the available sites. If this occurs the catalyst is said to be product poisoned. Rather these products must readily diffuse out of the pores and away from the surface. The subject of diffusion was treated earlier and will not be repeated here.

The overall catalyst reaction rate depends upon the rates of the five independent steps discussed above. Usually one of the steps will be slower than the rest and will be rate controlling. There is no reason that the overall rate will display Arrhenius-type behavior since only one of the five steps behaves in this way. Instead the overall rate may well be limited by mass transport and not by chemical kinetics, in which case many factors other than reactant concentration and temperature will be critical. Factors such as the nature of the catalyst surface, the size of its pores, if any, and its geometric arrangement with respect to the flowing gas all may have an important bearing on the rate of mass transport and hence the overall rate of catalytic reactions. These factors will be explored in the following section.

Nature of the Catalyst Surface

The rate of a catalytic reaction is proportional to the interfacial area on which the reaction takes place. If the catalytic material is metallic it may be used in the form of a wool, a screen, or sintered disc as illustrated in the upper portion of Figure 43. All of these shapes expose a relatively large amount of surface per unit mass of metal. In some applications the physical properties or the cost of the metallic material are such that it is not desirable to use the catalyst in its pure form. Instead, a fine coating of the metal may be dispersed on an inert surface. The catalyst is then referred to as a supported catalyst. The inert material may be a ceramic in the form of small balls, cylinders or a monolithic honeycomb structure. Two configurations of supported catalysts are illustrated in the lower portion of Figure 43.

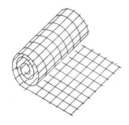

METALLIC CATALYSTS

SUPPORTED CATALYSTS

Figure 43. Catalyst configurations.

When the catalytic material is a nonstoichiometric oxide such as M_xO_y introduced earlier in Figure 42, it may be fabricated directly into the desired geometric shape (rods, balls, monolithic shapes).

While the geometric shapes illustrated in Figure 42 do present a large interfacial area to the flowing gases, this area may be further increased by the use of material that is porous and whose internal surface is catalytically active, that is, a material containing numerous internal passages (pores) into which the reactants can diffuse and reaction can take place on the internal surface area. Figure 44 portrays a porous sphere.

These internal pores vary in size from very small ones with a diameter of only a few microns to much larger ones. The pore size and size distribution of a catalyst can have an important effect on its operation.

All of the actual internal surface may not actually be utilized. If as reactant molecules diffuse in the pores they collide with the interior catalyst wall and undergo reaction before penetrating into the depths of the pores, all reaction may occur in a thin shell near the catalyst surface. Only a fraction of the total available area is actually utilized. This is referred to as the eggshell effect and is illustrated in Figure 45. The pores penetrate the entire volume of the catalyst; however, reaction

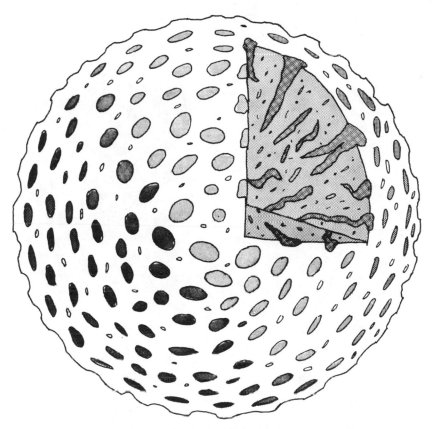

Figure 44. Cutaway of porous sphere showing internal pores.

occurs before the reactants have diffused very far from the external surface so only a thin eggshell-like region at the periphery of the cata- lyst actually catalyzes the reaction. This unused region is dotted in Figure 45. This effect is particularly likely to occur when the space velocity is high.

Frequently catalysts are prepared by utilizing an impervious ceramic to which a highly porous washcoat has been applied. The catalyst is then applied to the washcoat. This provides an actual eggshell of catalyst while at the same time incorporating the superior physical strength of the impervious support material.

Catalyst Environment

The catalyst environment influences the performance of a catalyst. Important factors include temperature, gas composition, (including any "poisons" that may be present) and flow-related factors (gas

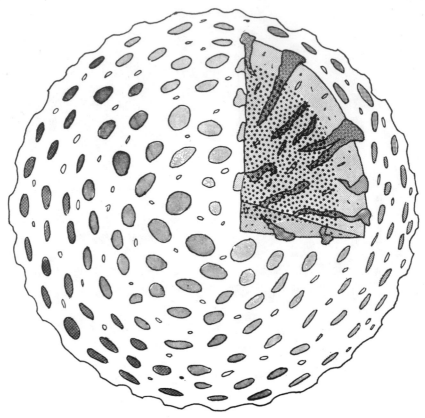

Figure 45. Cutaway of catalyst particle illustrating the "egg-shell" phenomenon.

velocity over catalyst surface, space velocity, turbulence, and pulsations). A number of the more important factors are discussed below.

Temperature

The minimum temperature for a catalytic reaction is usually determined by one of two considerations. Either it is the minimum temperature at which an acceptable degree of conversion can be attained, or it is the temperature below which the catalyst promotes the wrong reaction (*i.e.*, produces an undesired product distribution).

The maximum operating temperature is commonly governed by material limitations.[4] At elevated temperatures the mobility of atoms in the catalyst, the washcoat and the substrate all increase. The small pores decrease as the material coalesces. Coalescence is a thermodynamically favored process because it reduces the surface free energy, which is large for a highly porous substance. This is accomplished by a significant reduction in the surface area on which the catalyzed reaction occurs and a corresponding reduction in the activity of the catalyst. The decrease in surface area is irreversible.

Another mechanism by which a porous ceramic substrate or washcoat may lose surface area at elevated temperatures is a change in its crystalline structure. For example, ɤ-alumina, which is highly porous, undergoes a series of irreversible phase transformations as it is heated from approximately 1400°F to 1800°F. The material is transformed from a highly porous structure to a dense and impervious one, α-alumina. This type of transformation may cause a slight decrease in total volume of the material and a drastic reduction in its specific surface area.

In addition to the mechanisms described above, when some metallic catalytic materials are dispersed on a substrate as depicted in the lower portion of Figure 43 exposure of this material to too high a temperature may cause it to agglomerate, coalesce or otherwise change its crystalline structure. This may destroy the catalytic properties of the material altogether. Consequently, there is a "window" with respect to catalyst operating temperature. If the temperature is either too low or too high, unsatisfactory results occur.

Gas Composition

As in the case of thermal reactors there are no flammability limit restrictions on catalytic reactions. Catalysts may be used to promote the reaction of very high concentrations of reactant species as well as for very low and even trace concentrations. Concentration does affect rate and in general an increase in concentration will result in an increase in the rate though the two are not necessarily proportional to one another.

There are two ways in which the composition of the reactant stream may have an important bearing on the performance of a catalyst. The first is the effect on product equilibrium. This is important because a catalyst only facilitates the approach to equilibrium. Relatively large concentrations of H_2, CO_2 and H_2O are present in the effluent of most combustion processes. These species may compete with other reactions on catalytic surfaces that promote the water-gas shift reaction.

The second factor in the composition of the reactant stream is the presence of "poisons." A poison is limited to some incidental substance, usually present in small amounts, that interfers with the desired operation of the catalyst. The phenomenon of product-poisoning wherein the product molecule is tenaciously bound to the catalyst and blocks all available reaction sites was mentioned in an earlier section. The catalyst bed may act as a filter and remove particulate material such as fly ash and carbonaceous particles from the combustion products. This material in turn can coat the catalyst surfaces and act as a physical barrier thus preventing the diffusion of reactants to the surface and into the pores. Metallic vapors present in the combustion products, particularly volatile substances such as mercury and lead and compounds of these substances, may condense on the catalyst surfaces. Some metallic vapors may destroy catalyst activity by combining with metallic catalyst materials to produce an alloy that is not a catalyst itself or they may simply coat the catalyst rendering the surface inaccessible. Various other substances such as silicones, chlorofluorohydrocarbons and phosphorous compounds that are present in the fuel may oxidize in the primary combustion process to produce substances that attack and corrode catalyst materials.

Flow Related Factors

High catalyst efficiency requires that the reactant gas flow be uniformly distributed through the bed so that large quantities of this gas do not channel through passages in or around the catalyst and thereby avoid being brought into intimate contact with the active surfaces. The velocity of the bulk gases over the catalyst surfaces can also be important since it influences the thickness of the boundary layer film through which reactant and product diffusion must occur.

Other factors such as whether the flow is steady state or pulsating are important. Pulsation can cause a catalyst bed composed of discrete particles (spheres, etc.) to impact with one another and lead to attrition of the washcoat, or of the substrate itself. When the catalyst material, which may be only a few atoms thick, is dispersed on the substrate surface, abrasive action caused by movement of the bed may remove it. These same concerns are relevant when the volumetric flow rate of gases through the catalyst bed is sufficiently high that it will lead to fluidization of a particulate bed. When pulsation or fluidization are likely, a monolithic-type catalyst or support may be preferred over a bed of discrete particles.

Pressure drop is often a factor when catalysts are employed to treat postflame gases. The high space velocities associated with this type of application require catalyst beds of relatively large particles or of monolithic structure to prevent excessively large pressure drops.

Miscellaneous Factors

Each application of catalysts presents its unique environmental conditions. When the catalyst is used intermittently, moisture may condense in the bed when it is not in operation. If the ambient conditions drop below the freezing point (32°F), this water may freeze in the bed and subject it to unusual physical stress due to its expansion on freezing. Subsequent heating of the catalyst may tend to flash this ice, producing additional stresses.

Other transient considerations that may be present in some applications include thermal shocks introduced by rapid heating and cooling of the catalyst bed.

Oxidation Catalysts

Various materials can catalyze the oxidation of the CO, H_2 and hydrocarbons present in the postflame gases. The basic arrangement for catalytic oxidation is shown in Figure 30. If the postflame gases do not already contain enough oxygen, sufficient secondary air must be added to oxidize all of the combustibles. There is no other limitation on the minimum amount of oxygen required. The flow scheme is not unlike that for the thermal reactor; however, in the case of the catalytic reactor lower temperatures can be used. Therefore the potential for promoting pyrolytic reactions under poor mixing conditions is reduced.

With the exception of the noble metals oxidation catalysts are invariably nonstoichiometric oxides and can gain or lose oxygen from their lattice structure. Typical nonstoichiometric oxides that promote oxidation are Co_3O_4, $CoO \cdot Cr_2O_3$, MnO_2, $LaCoO_2$, CuO, Fe_2O_3, Cr_2O_3, NiO & V_2O_5.[5] By contrast the stoichiometric oxides Al_2O_3 and SiO_2 do not by themselves exhibit catalytic activity with regard to oxidizing combustibles. Due to their superior physical properties these stoichiometric oxides are often employed as substrates to support catalysts.

A mechanism for CO oxidation by a nonstoichiometric oxide was presented earlier in Figure 42. By way of contrast, the oxidation of hydrocarbons on a catalyst surface involves an even more complex series of steps that differ for each particular type of hydrocarbon molecule. In general the reactivity or ease of oxidation of hydrocarbons depends upon molecular structure. The concept of catalytic reactivity is a complex one and can be understood by examining Figure 46. This figure shows the relationship between fractional conversion and temperature for the catalytic oxidation of two hydrocarbon species A and B. A is more reactive than B since a specified degree of conversion (i.e., $F_c = 0.5$) occurs at a lower temperature for species A than for species B. At temperature C the conversion of A is nearly complete, while only a small fraction of B has reacted. When the catalyst is operated at this temperature it is said to be selective with respect to the oxidation A in a mixture of A and B. As the operating temperature increases, the selectivity disappears. If different types of hydrocarbon molecules with

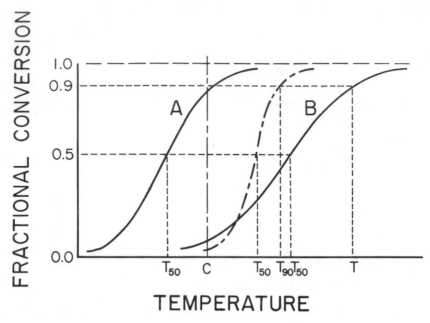

Figure 46. Relationship between fractional conversion and temperature for the catalytic oxidation of hydrocarbons.

the same number of carbon atoms per molecule (*i.e.*, carbon number) are compared, acetylenic species are the most reactive, followed by olefins and then paraffins. Aromatics are the least reactive (or most refractory). If hydrocarbons with a specific group (*e.g.*, paraffins only or olefins only) are compared, reactivity decreases as the carbon number decreases. As expected, methane is the most difficult of all hydrocarbons to oxidize catalytically.

Reactivity also depends upon the particular catalyst. For example, in Figure 46 two separate curves are shown for species B. The broken curve is typical of the relationship between conversion and temperature when oxidation is promoted by a noble metal catalyst. The solid curve depicts the relationship for oxidation promoted by a nonstoichiometric oxide. In this latter case the oxidation begins at a lower temperature but a higher temperature is required for complete oxidation. Two temperatures are shown here: T_{50}, corresponding to 50% oxidation of the reactant species, and T_{90}, corresponding to 90% oxidation. These were selected because they are the most common choices for denoting *light-off temperature* or, in other words, the lowest temperature at which the catalyst has appreciable activity with regard to oxidizing the reactants. The definition of light-off temperature is arbitrary and other levels of conversion can be selected if desired.

It should also be noted that fractional conversion as defined does not imply complete oxidation. It means that a certain fraction of a species that entered the reactor was converted to something else within the reactor. This something else may include partially oxidized species as well as completely oxidized ones. In this case, as is also true in the cases of the thermal reactor and the afterburner, the specification of a fractional conversion as defined in Equation (61) cannot be assumed to imply that the emission of undesired species is reduced accordingly. The product of the reaction is also of importance for it may be more or less desirable than the original species.

An oxidation catalyst is generally selected to remove undesirable species from the postflame gases. One particular side reaction that may occur, the production of partially oxidized species from unoxidized ones, has just been mentioned. The result is generally undesirable. Other types of side reactions can also occur. Generally a good oxidation catalyst for CO and hydrocarbons is also a good oxidation catalyst for other species, such as sulfur and its oxides. When the fuel used in primary combustion processes contains sulfur, this sulfur will be converted almost entirely into sulfur dioxide within the flame. As the postflame gases cool, the equilibrium of the following reaction shifts toward the right.

$$2SO_2 + O_2 \rightleftharpoons 2SO_3 \tag{74}$$

However, the rate of this reaction decreases so rapidly with a decrease in gas temperature that little (generally only a few per cent) of the total sulfur is present as SO_3. The bulk of the sulfur is "frozen" in the postflame gases as SO_2. An oxidation catalyst for the reaction will promote the rate at which equilibrium is attained and increase the amount of SO_3 in the effluent at the expense of SO_2. Neither SO_2 nor SO_3 is a desirable species in the effluent. The catalytic promotion of Reaction (74) is not desirable if for no other reason than when the effluent gases cool even further, the SO_3 present will increase the acid dew point and lead to the formation of an acid mist that can corrode surfaces in the exhaust system. At the lower operating temperatures, the SO_3 may react directly with the catalyst material or the substrate to produce sulfates. These compounds may subsequently flake off, resulting in particulate sulfate emissions.

In a like manner oxidation catalysts will promote the oxidation of still other gaseous species that can be oxidized. For example, if ammonia is present it will be oxidized to nitric oxide.

$$2\,NH_3 + 5/2\,O_2 \rightleftharpoons 2\,NO + 3\,H_2O \qquad (75)$$

The presence of carbonaceous particles in the postflame gases presents special problems. First of all, the catalyst may act as a filter and remove these particles from the gases. This is true for particles of any composition, carbonaceous or otherwise.

When the particles are carbonaceous they may undergo subsequent thermal oxidation (not catalytic) as they are attached to the catalyst surface and exposed to the hot flowing gases. If there is sufficient oxygen in these gases, the particles may oxidize. Due to the low thermal conductivity of many catalyst substrates the energy liberated by these oxidizing particles can result in localized "hot-spots" on the catalyst surface. This in turn can result in local catalytic degradation.

Proper functioning of an oxidation catalyst requires that it remain within a certain operating temperature "window." If the inlet gases are at too low a temperature, a preheater arrangement (see Figure 37) similar to that used with an afterburner can be used. Too low a temperature may be a problem when a catalyst is being employed to oxidize species in an effluent stream from a noncombustion source. This is the reverse of what usually exists when catalysts are used to clean up the effluent from combustion systems. The postflame gases from a combustion process are hot, sometimes too hot to pass directly into the catalyst without some precooling. To compound the problem, the oxidation of combustibles is exothermic. The quantity of energy liberated by a reaction is independent of the path, *i.e.*, whether it is homogeneous or heterogeneous. Thus the adiabatic temperature rise in the catalyst bed can be estimated from Figure 29. To prevent the rise in temperature from destroying the catalyst special precautions must be taken. First of all the reactor itself can be designed so that it is not adiabatic and the temperature rise will be less than that indicated in Figure 29. Second, the catalyst bed can be broken into two stages and an intercooler placed between the two. This approach is illustrated in the top portion of Figure 47. An alternate approach to temperature control is to divide the secondary air between the two stages as is shown in the bottom part of the figure. This approach is particularly useful when the temperature and/or the concentration of combustibles in the postflame gases may undergo fluctuations. By limiting the oxygen concentration in the first stage the maximum possible temperature rise is also limited.

Reduction Catalysts

Reactions (68), (69) and (70) may be promoted by noble metals (Pd or Ru), by base metals such as Co, Ni and Cu and by combinations

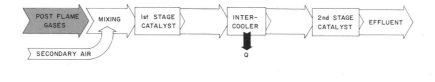

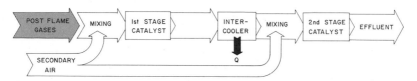

Figure 47. Two-stage catalytic oxidation.

thereof. Some oxides, for example CuO and $CuCrO_4$, also promote the catalytic reduction of NO.[6,7] The selection of a catalyst for a particular application depends upon a variety of factors. The noble metals are costly and are usually applied to ceramic substrates. They promote catalytic reactions at relatively low temperatures. Base metals may be used in unsupported forms and usually require higher temperatures to accomplish the same degree of reduction. Among the other important considerations in the selection of a catalyst are the products formed and the requirements of the particular catalyst with respect to the maximum amount of oxidant and/or reductants which can be tolerated. These factors will be discussed later.

In promoting the reduction of NO, the object is to obtain N_2 as shown in Reactions (68), (69) and (70). However, as is the case with most catalytic reactions, a number of products can be formed other than the desired one(s). Reaction (68) occurs via a two step mechanism as shown below. At low temperatures the rate of Reaction (77) is slow compared to that of (76), the reduction may be partial, and the nitric oxide is only reduced to nitrous oxide.[6,7]

PARTIAL $\qquad 2NO + CO \rightleftharpoons N_2O + CO_2 \qquad (76)$
$\underline{\qquad\qquad\qquad N_2O + CO \rightleftharpoons N_2 + CO_2 \qquad (77)}$
COMPLETE $\qquad 2NO + 2CO \rightarrow N_2 + 2CO_2 \qquad (68)$

The relationship between catalyst temperature and the product distribution (mix of N_2O and N_2) for a particular catalyst is shown in Figure 48. At low temperatures the bulk of the NO is converted to N_2O. As the temperature rises, the rate of Reaction (77) increases and more of the NO is completely reduced to N_2.

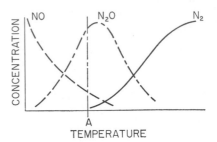

Figure 48. Relationship between product distribution and temperature for a particular reduction catalyst.

A portion of the NO converted by the reducing catalyst reacts with hydrogen to form ammonia:[8,9]

$$\frac{2}{5}NO + H_2 \rightarrow \frac{2}{5}H_2O + \frac{2}{5}NH_3 \qquad (78)$$

There are two sources of hydrogen for this reaction. The first is molecular hydrogen present in the postflame gases. Examination of Figures 1 and 21 reveals that on a mol basis the concentration of hydrogen is from $\frac{1}{4}$ to $\frac{1}{3}$ that of carbon monoxide. A second source of hydrogen is the water-gas shift reaction. Reduction catalysts generally absorb water vapor and promote the water-gas shift reaction.

$$H_2O + CO \rightleftharpoons H_2 + CO_2 \qquad (47)$$

The formation of ammonia depends upon a variety of factors. First is the composition of the catalyst itself. Platinum may convert as much as 50% of the NO to ammonia while other materials, notably those containing ruthenium, produce little or no ammonia. The operating temperature of the catalyst is also a factor. Figure 49 illustrates how the product distribution produced by one platinum catalyst varied with temperature. At low temperatures the gross conversion is relatively low; however, all of the NO is converted to molecular nitrogen. Beginning at approximately 800°F an appreciable fraction of the NO is converted into NH_3. The concentration of reductants in the gases also influences the product distribution. In general ammonia formation increases as the concentration of H_2 and CO increases. This is illustrated in Figure 50. This type of behavior is not unexpected since the supply of H_2 for Reaction (78) increases due both to the greater amount of hydrogen produced by the primary combustion process and the increased production of hydrogen on the catalyst surface itself due to the water-gas shift reaction.

With the possible exception of nickel, good reducing catalysts are also good oxidizing catalysts. So good in fact, that if oxygen is present

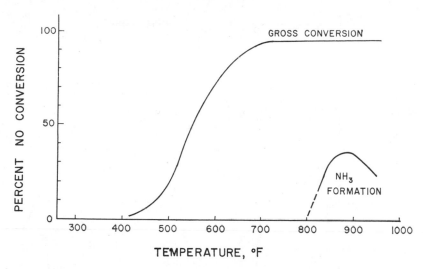

Figure 49. Relationship between temperature and product distribution for the catalytic reduction of NO on a particular platinum catalyst.

the reductants H_2 and CO are oxidized before they can reduce the NO. Thus, for reduction to occur the gas entering the catalyst must be net reducing, that is, residual oxygen can be present; however, the concentration of reductants (HC & CO) must be sufficiently large to completely consume this oxygen and still have enough reducing species remaining to react with the NO.

From the above discussion it is apparent that a significant number of very different reactions can be promoted by a so-called "reduction catalyst." These possibilities are summarized in Figure 51. Promotion of the desired product distribution requires careful regulation of the catalyst operating temperature and of the composition of the gases entering the catalyst.

In addition to the major reactions discussed above a number of unwanted reactions may also occur. First, in a reducing atmosphere, hydrocarbons may crack. Consequently the concentration of simple species like methane may increase at the expense of more complex hydrocarbon species. A second area of concern is the effect of a reducing catalyst on particles. As with an oxidizing catalyst, reducing catalysts may filter particles from the flowing gases. However, unlike the situation that exists with an oxidizing catalyst there is usually insufficient oxygen present to remove these particles. If they undergo any reaction at all, the nature of such reactions is likely to be pyrolytic rather than oxidative.

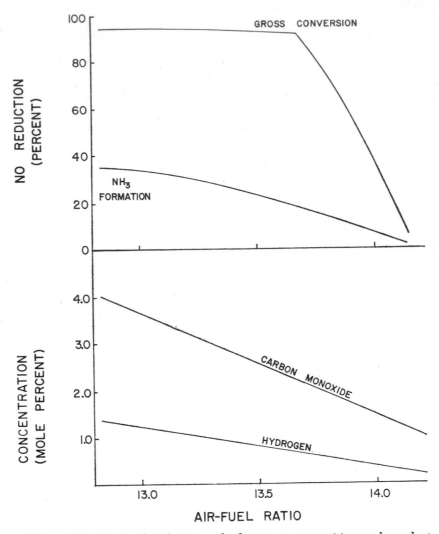

Figure 50. Relationship between feed stream composition and product distribution for the catalytic reduction of NO on a particular platinum catalyst.

Finally, the reduction of NO is an exothermic reaction. For example, approximately 180 kcal/mol are liberated by Reaction (68). In combustion systems the total NO is rarely more than a few tenths of a per cent. Consequently, even though the heat of reaction per mole of NO is large, the total heat liberated per mole of gas treated is usually small and the catalytic reduction can be thought of as an only slightly exothermic process.

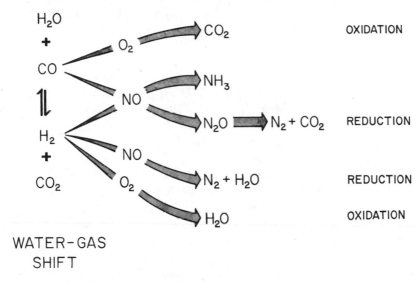

WATER-GAS
SHIFT

Figure 51. Competing reactions on a reduction catalyst.

Multiple-Bed Catalyst Systems

Thus far the discussion of catalysts has focused either on their ability to oxidize combustibles or to reduce nitric oxide. In some situations the oxidation and reduction reactions may be accomplished simultaneously. If the mole ratio of total reductants, (*i.e.*, combustibles $H_2 + CO + HC's$) to nitric oxide is such that complete reaction of all species according to Reactions (68), (69) and (70) is possible, then a single catalyst is theoretically capable of accomplishing the simultaneous elimination of all the unwanted species. The closest that this optimum is approached in practice is with the so-called three-way catalyst. This type of catalyst successfully reduces the concentrations of CO, H_2, hydrocarbons and nitric oxide when the primary air-fuel ratio is maintained very close to the stoichiometric value. Here the relative quantities of nitric oxide and combustibles are such that Reactions (68), (69) and (70) produce large decreases in the concentrations of both. The degree of precision that must be maintained with respect to the primary air-fuel ratio in order to insure the proper ratio of reactants in the fuel to the catalyst is so stringent that a three-way catalyst is not practical in many applications. If the air-fuel ratio of the primary combustion process is only very slightly too lean, the combustibles will be oxidized as per Reactions (65), (66) and (67) and no reduction of the nitric reaction will occur. Conversely, if the air-fuel ratio is slightly too rich, reduction of the nitric oxide will occur but the amount of combustibles entering the catalyst will be so great that not all of them can be consumed by Reactions (68), (69) and (70). The unreacted combustibles will be present in the effluent.

An alternate approach for simultaneous catalytic control of nitric oxide and combustibles is to place a reduction catalyst in series with and before an oxidation catalyst.[10] The air-fuel ratio of the primary combustion process is adjusted to be on the rich side of stoichiometric. This produces reductants that, when passed through the reduction catalyst, react with any nitric oxide formed in the primary combustion process. Secondary air is added between the two catalysts and excess combustibles are consumed by the oxidation catalyst. This arrangement is illustrated at the bottom of Figure 30. Note that an initial oxidation step is shown as preceding the reduction catalyst. This is required only when the primary combustion effluent contains appreciable oxygen. This oxidation step may be thermal or catalytic in nature. Sufficient secondary fuel must be added to insure net reducing conditions at the inlet to the reduction catalyst. This additional step is unnecessary if the influent to the reduction catalyst does not contain oxygen.

The primary advantage of this multiple catalyst approach as compared to the three-way catalyst is that the use of multiple catalysts requires less precise control over the air-fuel ratio of the primary combustion process. At least small excursions of the air-fuel ratio from the optimum can be compensated for by one or the other of the catalysts. For example, if the air-fuel ratio is slightly richer than desired, reduction of nitric oxide will proceed as usual while the excess combustibles will be consumed by the oxidation catalyst, as long as there is an overall excess of air at this point. Conversely, if the air-fuel ratio is not as rich as desired, reduction will continue as long as it remains on the rich side of stoichiometric.

However, there is a limit to the excursions in air-fuel ratio that a multiple-bed catalyst can tolerate. On one hand the primary air-fuel ratio must not become so lean that the inlet to the reducing catalyst is no longer net reducing. On the other hand there is a limit to the degree of acceptable richness. This will be apparent from an examination of Figure 50. As the air-fuel ratio decreases, more and more of the nitric oxide is reduced to ammonia per Reaction (78) instead of to molecular nitrogen per Reactions (68), (69) and (70). This ammonia will subsequently be reoxidized to nitric oxide (Equation 75) in the oxidation catalyst. Therefore, though the gross conversion of nitric oxide in the reduction catalyst may be great, as shown in Figure 50, the net conversion of a system composed of a reduction and oxidation catalyst in series will be less. This consequence is illustrated in Figure 52, which shows the net conversion that results when an oxidation catalyst is used to treat the effluent from the reduction catalyst of Figure 50.

Similar reasoning applies to the operating temperature of the reducing catalyst. At low temperatures the gross conversion is low, as shown in Figure 49. At high temperatures an appreciable fraction of the

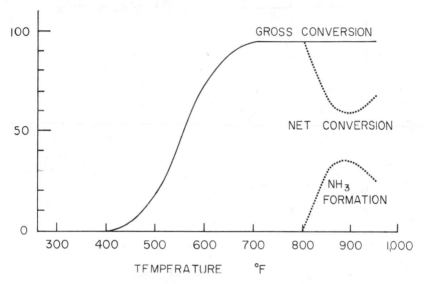

Figure 52. Net conversion achieved with a particular oxidation and reduction catalyst in series.

nitric oxide is converted to form ammonia, which when passed through an oxidation catalyst will be oxidized to nitric oxide. The net conversion corresponding to the case first illustrated in Figure 49 is presented in Figure 53.

From the above discussion it should be apparent that even with a multiple-bed catalyst system, a degree of control must be exercised over operating variables. Figure 54 shows the surface of net conversion that is achieved for a reduction and an oxidation catalyst in series when the reduction catalyst temperature and primary combustion air-fuel ratio are allowed to vary. The degree of control required depends upon the net conversion desired. For the case shown, in order to achieve 90% net reduction, which corresponds to the shaded area at the peak of the surface, the temperature of the reduction catalyst must be maintained between 650 and 750°F. The air-fuel ratio of the primary combustion process must also be maintained between 12.9 and 13.6. This is not inconsequential for it represents a ±3% control of the air-fuel ratio.

It is less than the degree of control that would be required to achieve a similar result with a three-way catalyst. The range shown here is presented only for illustrative purposes. It must be stressed that the concept of a range, or a window as it was previously referred to, is a

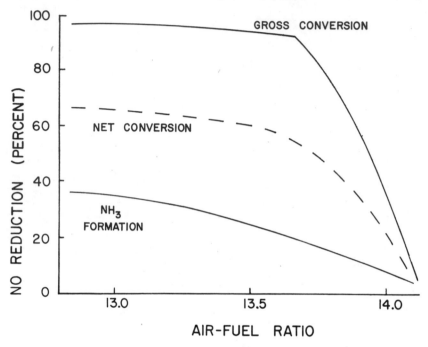

Figure 53. Net conversion achieved with a particular oxidation and reduction catalyst in series.

property of a particular catalyst or group of catalysts. Technological developments may yield catalysts that have wider ranges than those which are available today.

3.5 SUMMARY

Three types of combustion-related secondary treatment processes have been discussed in this chapter: the thermal reactor, the afterburner, and the catalytic reactor. The inclusion of this subject matter after the discussion of premixed flames and before diffusion flames should not be construed to mean that these processes are only applicable for the treatment of effluent from premixed flames.

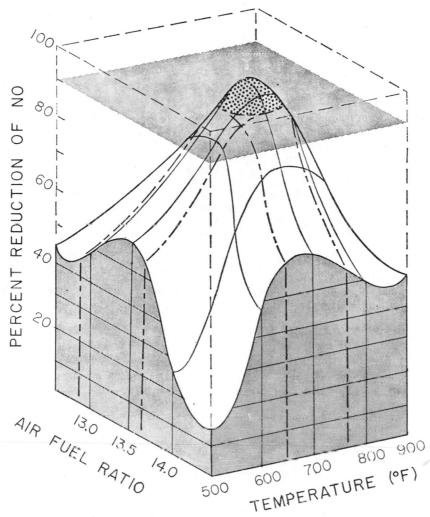

Figure 54. Net reduction of NO achieved with a particular oxidation and reduction catalyst in series as a function of the air-fuel ratio of the primary combustion process and the temperature prevailing in the reduction catalyst.

The selection of a secondary combustion process depends upon the particular trace elements present in the postflame gases that must be reacted rather than upon the nature of the primary combustion process. It will become readily apparent that the secondary combustion processes discussed in this chapter have considerable potential for the treatment of the effluent of diffusion as well as premixed flames.

REFERENCES

1. Levenspiel, O. *Chemical Reaction Engineering* (New York: John Wiley, 1962).
2. Vulis, L. A. *Thermal Regimes of Combustion,* translated by M. D. Friedman (New York: McGraw-Hill Book Co., 1961).
3. Stahl, Q. R. "Preliminary Air Pollution Survey of Aldehydes–A Literature Review," National Air Pollution Control Administration Publication No. APTD 69-24 (October 1969).
4. Farrauto, R. J., R. D. Shoup, and K. E. Hoekstra. "Deterioration of an Auto Exhaust Catalyst," Central States Meeting of the Combustion Institute, Champaign, Ill. (March 27-28, 1973).
5. Thomas, J. M., and W. J. Thomas. *Introduction to the Principles of Heterogeneous Catalysis* (New York: Academic Press, 1967).
6. Baker, R. A., and R. C. Doerr. "Catalyzed Nitric Oxide Reduction with Carbon Monoxide," *Ind. Eng. Chem., Process Des. Develop.,* **4,** 188 (April 1965).
7. Bauerle, G. L., G. R. Service, and K. Nobe. "Catalytic Reduction of Nitric Oxide with Carbon Monoxide," *Ind. Eng. Chem., Prod. Res. Develop.,* **11,** 54 (1972).
8. Shelef, M., and H. S. Gandhi. "Ammonia Formation in Catalytic Reduction of Nitric Oxide by Molecular Hydrogen," *Ind. Eng. Chem., Prod. Res. Develop.,* **11,** 2 (1972).
9. Jones, J. H., J. T. Kummer, O. Klaus, M. Shelef, and E. E. Weaver. "Selective Catalytic Reaction of Hydrogen with Nitric Oxide in the Presence of Oxygen," *Environ. Sci. Technol.,* **5,** 790 (1971).
10. Shelef, M., K. Otto, and H. Gandhi. "The Oxidation of CO by O_2 and by NO on Supported Chromium Oxide and Other Metal Oxide Catalysts," *J. Catal.,* 361 (1968).

DIFFUSION
FLAMES

4.1 INTRODUCTION

Premixed combustion was discussed in Chapter 2. In that chapter the term *premixed* was reserved for the case in which intimate mixing of the fuel and the air exists on a molecular scale. Combustion in premixed flames is usually limited to fuels that are gaseous at ambient temperatures or to those that vaporize at relatively low temperatures. There are many practical combustion systems in which the fuel and the air are not premixed on a molecular scale. Rather, the fuel and the air remain separated until brought into intimate contact with one another in the immediate vicinity of the flame itself. A complex set of processes occur in the flame including mixing, precombustion, combustion and postflame reactions and vaporization when the fuel is a liquid or volatilization when it is a solid. The term *diffusion flame* is used to denote this type of combustion process, *i.e.*, one in which the fuel and the air are not premixed.

It was also noted in the introduction to Chapter 2 that with the exception of the conventional automobile engine there are relatively few practical applications of premixed combustion. The great bulk of actual combustion devices burn their fuel in a diffusion flame, partly for safety reasons and partly for other considerations. Whatever the reasons, the point to be made here is that the emissions from diffusion flames are important from an air pollution viewpoint.

One of the things that will be apparent to the reader is the diverse kinds of diffusion flames. Perhaps the most familiar diffusion flame is that occurring on the burners of domestic stoves. This is true for both natural gas and LPG fuels. A quite different type of diffusion flame occurs in a Diesel engine. Here the fuel is sprayed into the combustion chamber as a fine mist, and individual diffusion flames are established about each fuel droplet. Other common combustion systems that consume fuels in a diffusion flame are the gas turbines that power jet aircraft, the burning of coal and the combustion of wood in a campfire.

The subject of propellant, monopropellant and explosives combustion is treated briefly at the end of this chapter. Recall that in Chapter 1 these systems were identified as premixed combustion systems. This is correct. The rationale for discussing them at this point is that there is a great similarity between the mechanisms by which these fuels burn and the combustion of solid fuels in general.

4.2 PRINCIPLES OF DIFFUSION FLAME COMBUSTION

There are a number of fundamental differences between diffusion flames and a premixed flame. Perhaps the differences can be best understood by observing the contrast between the two. The sequential arrangement of precombustion, combustion and postflame processes in a premixed flame was first presented in Figure 3. By way of contrast the arrangement of these same processes in a diffusion flame is shown in Figure 55.

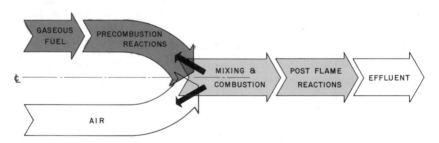

Figure 55. Diffusion flame.

One of the fundamental differences is that in a diffusion flame the precombustion reactions occur *before* mixing of fuel and air. In the premixed flame, the precombustion reactions occur after intimate mixing of the fuel and air. Thus, the environments in which the pre-combustion reactions take place for the two types of flames differ chemically. The precombustion zone of a diffusion flame does not contain oxidizing species ($O^{\bullet}$, $^{\bullet}OH$, O_2, etc.) and reactions analogous to (9) through (16), which occur in the precombustion zone of a pre-mixed flame, cannot occur in the precombustion zone of the diffusion flame. Because the precombustion zone of a premixed flame can contain only fuel molecules or fragments thereof, the environment for pre-combustion reactions is reducing in character. Under these conditions precombustion reactions are expected to be similar to pyrolysis, such as represented earlier in Equations (48) through (54). The driving forces for these pyrolytic precombustion reactions are the countercurrent transport of thermal energy and active species from the combustion zone to the precombustion, as denoted by the upper solid arrow in Figure 55.

Because the precombustion reactions of a diffusion flame are pyrolytic, the reaction products include unsaturated species such as olefins and acetylenes and particulate nuclei resulting from polymerization or addition reactions between these unsaturated species (refer to the earlier discussion in the section on postflame reactions). The net result of the appearance of particulate species in the precombustion zone is to enhance the thermal coupling between the precombustion and combustion zones of a diffusion flame. When these particles pass through the flame, they radiate as black bodies (Figure 12). This is responsible not only for the characteristic luminosity of diffusion flames but also for the transfer of radiant energy to the particle nuclei forming in the precombustion zone. Since particles are good adsorbers as well as good emitters of radiant energy, the particles forming in the precombustion zone absorb the energy radiated from the combustion zone. A portion of this absorbed energy is transferred by conduction to gaseous species in the precombustion zone, which in turn increases their temperature and promotes gas phase pyrolytic reactions. Thus for a diffusion flame the exchange of radiant energy by particles that are formed in the precombustion zone and later pass through the flame plays an important role in the energetics and chemistry of the flame. In fact, it is this coupling of zones that drives the pyrolytic precombustion reactions that are generally endothermic. In the case of the premixed flame, coupling of the combustion and precombustion zones does occur; however, because the oxidative reactions in the precombustion zone are exothermic, they can lead to auto ignition in the absence of a flame.

There is another basic dissimilarity between premixed and diffusion flames. In the case of the premixed flame, sustained ignition takes place when the quantity of thermal energy liberated by the oxidative precombustion reactions is sufficient to overcome thermal losses from the reacting mixture and raise it to its ignition point. This same explanation is not sufficient for describing the ignition process in a diffusion flame. In this case, as the fuel is initially separated from the oxidant, combustion, *i.e.*, oxidation of the fuel, cannot occur regardless of how high the temperature might rise before mixing of the fuel and oxidant produces a combustible mixture. Thus ignition in a diffusion flame is controlled by the physical processes that influence mixing such as turbulence and the geometry of the system. By contrast, ignition in a

premixed flame is controlled by chemical kinetic processes that influence the rate and extent of heat release and the degree of chain
branching.

One additional observation with respect to the influence of mixing on
the reaction products of a diffusion flame. Because oxidation cannot
occur until fuel and oxidant are mixed, the pyrolytic precombustion reactions will continue until mixing does occur. If final mixing is poor,
these pyrolytic reactions may produce relatively large quantities of
particulate material and other pyrolytic reaction species. From this perspective it is easy to understand the importance of the mixing step and
the propensity for diffusion flames to form copious amounts of carbonaceous particulate matter when this mixing step is inefficient.

The upper portion of Figure 56 depicts a diffusion flame. The fuel
and oxidant are separated and the flame occurs in an interfacial zone
where the two meet and mix. Radiant energy transferred from the flame
preheats the air and provides the driving force for the pyrolytic precombustion reactions that occur as the fuel approaches the flame
zone. The precombustion zone is not specifically identified in this
figure. The lower portion of the figure depicts the relationship of fuel
and oxidant on either side of and within the diffusion flame. The flame
itself has well-defined boundaries. To the left of the flame the ratio
of fuel to oxidant is too large for combustion to be sustained. Thus the
left boundary in Figure 56 is analogous to the rich flammability limit
discussed in Figure 19. In a similar fashion the right-hand limit corresponds to the lean flammability limit; to the right of this there is
insufficient fuel to sustain combustion. The absolute width of the combustion zone will depend upon a number of factors including the
nature of the mixing process that brings the fuel and the oxidant into
intimate molecular contact.

Within the combustion zone the ratio of fuel to oxidant (or fuel to air)
can be thought of as varying continuously from the rich to the lean
limit. The fuel and the oxidant do not react at any specific air-fuel
ratio but at a progressively changing one. This observation has two
ramifications of particular interest. First, it is not possible to exercise
control of the combustion by chosing a particular air-fuel ratio that is
favorable with respect to the reaction products desired. This, of course,
is one of the options available for controlling emissions from a premixed
flame. A second observation is that a diffusion flame is inherently more
stable than a premixed one. In the latter case if the overall or local
air-fuel ratio exceeds the rich or the lean flammability limits, combustion ceases with the usual consequence of high emissions of fuel or
partial oxidation products thereof. In the case of a diffusion flame, a

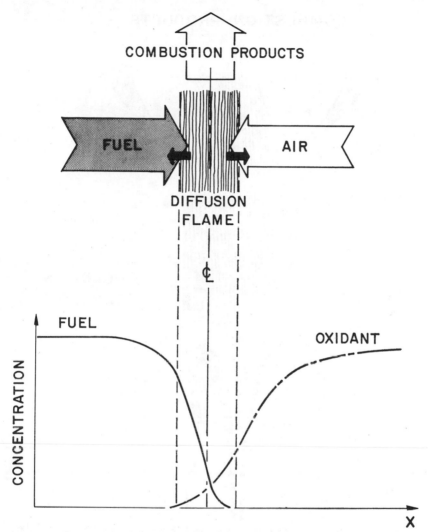

Figure 56. Spatial relationship of fuel and oxidant in a diffusion flame.

combustible mixture will always exist at some point between the fuel and the air unless the flow of one or the other of these components is totally interrupted. Transient upsets in the flow will not result in the flame being extinguished as easily.

All classes of fuels, gases, liquids and solids can be burned in a diffusion flame. Figure 57 illustrates the combustion of a gaseous fuel in such a flame. This figure is analogous to the picture of a Bunsen burner shown earlier in Figure 4 for the combustion of the same gaseous

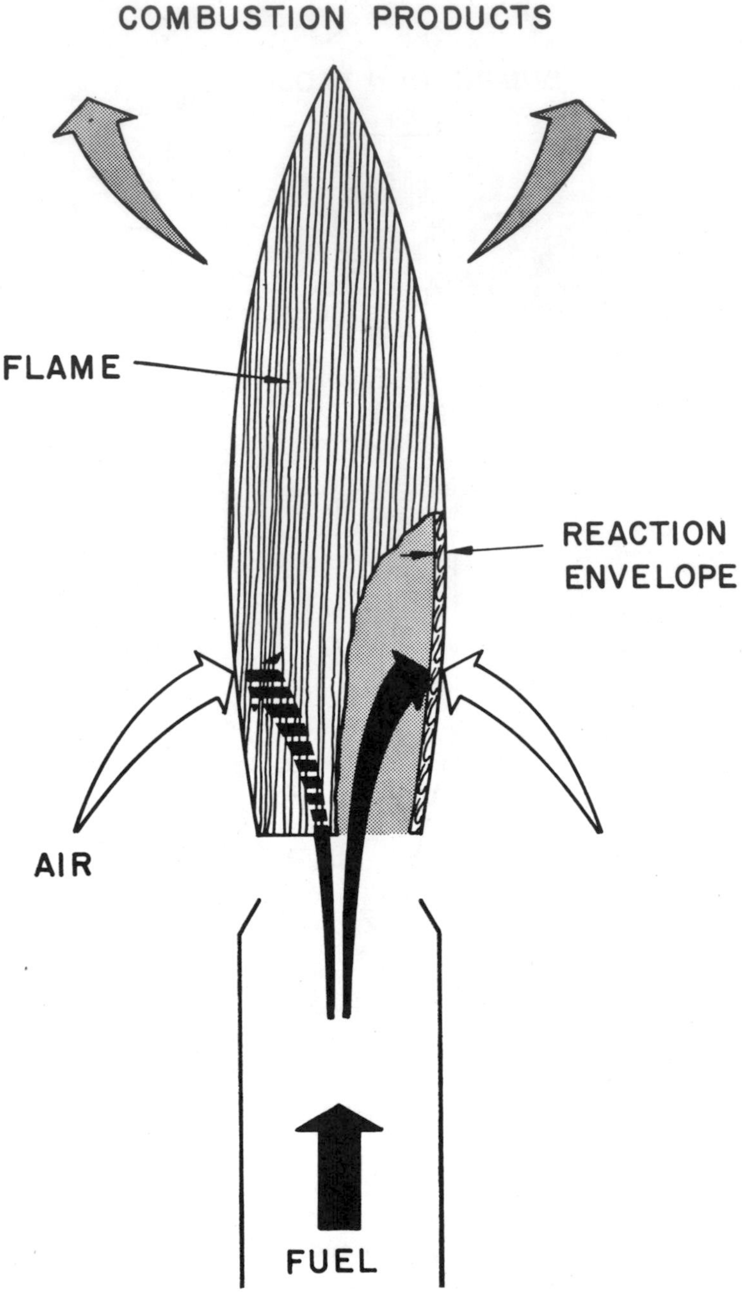

Figure 57. Combustion of a gaseous fuel in a diffusion flame.

fuel in a premixed flame. Note that in Figure 57 the fuel is shown as reacting in a thin reaction envelope (the flame). Precombustion reactions take place in the region within this envelope as the fuel passes from the neck of the jet to the reaction envelope. The air is drawn in around the base of the flame as the hot combustion products rise due to their buoyancy relative to the surrounding atmosphere.

The diffusion flame as shown in Figure 57 is inherently more stable than its premixed counterpart presented earlier in Figure 4. In the latter case, the velocity of the premixed reactants emanating from the Bunsen burner throat must be adjusted to be reasonably close to the burning velocity of the fuel-air mixture for the flame to remain stationary. If this is not accomplished and the velocity of reactants at the throat is too high, the flame will "blow off," while if it is too low "flash back" will occur as the flame propagates back into the mixing chamber. In the case of the diffusion flame flash back is not even a possibility since there is no air in the fuel supply, and when the gas velocity increases the length of the reaction zone can change to accommodate this without much likelihood of blow off.

As was mentioned earlier liquids and solids can also be burned in diffusion flames. In fact this is the rule rather than the exception. There are an almost unlimited number of ways in which liquids and solids can be mixed with air to promote combustion. Specific applications will not be discussed in detail. Only the general principles of liquid and solid fuel combustion will be considered.

4.3 COMBUSTION OF LIQUIDS

Vaporization

A liquid fuel does not burn as such. It must first be vaporized and the vapor mixed with surrounding air to form a combustible mixture. The basic processes associated with a diffusion flame were presented in Figure 55. When the fuel is initially a liquid rather than a gas, a vaporization step must be added as shown in Figure 58.

The vaporization process is of great interest in this discussion because the manner in which it is accomplished influences the combustion process and the products thereof. Since the vaporization and mixing steps are in series, the rate of vaporization may significantly influence and may even control the overall rate of combustion. There are a number of factors that influence the rate of vaporization of a fuel. The more important ones will be discussed.

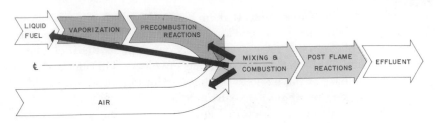

Figure 58. Combustion of a liquid fuel in a diffusion flame.

Dispersion of the Liquid

The rate of vaporization generally increases as the surface area of the liquid increases. Consequently to increase the rate of vaporization and thereby permit the achievement of greater combustion intensities the fuel is commonly broken up (atomized, etc.) in the form of a spray. For purposes of discussion large aggregates of fuel (diameter greater than 1 cm) will be referred to as pools. Droplets have diameters ranging from $10^4\mu$ down to $10^2\mu$. Fuel droplets smaller than this are termed mists.

Energetics of the Vaporization Process

Vaporization is an endothermic process, that is, energy must be added to a liquid fuel to convert it to a vapor. This energy is commonly referred to as the latent heat of vaporization. When the latent heat is supplied by cooling of the liquid and of the surrounding gases the process is usually referred to as evaporation. It is often a relatively slow process. One item of significance is that evaporation of a fuel decreases the temperature of the resultant air-fuel mixture, and consequently adiabatic flame temperature will be reduced also. The implications of this with respect to the formation of nitric oxide have been discussed previously.

Combustion processes often involve much more rapid conversion of liquid fuel to fuel vapor than is accomplished by simple evaporation. The energy to drive this rapid conversion of liquid to vapor is usually supplied by the combustion process itself. Energy radiated by particles passing through the diffusion flame is transmitted through the largely transparent fuel vapors and is absorbed by the liquid fuel.[1] This energy is denoted by the solid arrow connecting the combustion and liquid fuel regions in Figure 58.

When a pure fuel (one component) is vaporized by the rapid addition of energy its temperature increases until the boiling point is reached and vaporization then proceeds at that temperature till all of the fuel is converted to a vapor. Many actual liquid fuels are multicomponent rather than single component. That is, they contain two or more different chemical species. Those components with the highest vapor pressures evaporate more rapidly. The remaining liquid is enriched in the lower vapor pressure (higher boiling temperature) components. Consequently the liquid temperature does not remain constant during the vaporization process, as was the case for a single-component fuel. It increases as the vaporization proceeds. If the liquid phase temperature exceeds the decomposition temperature of some components remaining, liquid phase pyrolysis occurs. These reactions in the liquid phase can ultimately produce high molecular weight carbonaceous material that is resistant to subsequent oxidation and appears in the effluent as particulate matter.

Pressure

The pressure at which a liquid fuel is vaporized is also important. As the pressure increases, higher temperatures are required to vaporize the fuel. At these higher temperatures liquid phase pyrolytic (decomposition) reactions of the heavier fuel components are more probable.

The vaporization process is also influenced by whether or not the pressure of the air exceeds the critical pressure of one or more components of the fuel. Pressure is important from still another point of view. Dispersion of the fuel is often accomplished by application of a high hydrostatic pressure to force the liquid fuel through a nozzle and convert it to a spray consisting of small droplets. As the liquid emerges from the nozzle the pressure exerted on it drops almost discontinuously while its temperature remains constant or nearly so. A fraction of the more volatile components of the fuel "flash." The resulting fuel vapor mixes with the air and may play an important role in initiating the precombustion reactions.

Mixing

After the liquid fuel has been vaporized it must be mixed with air before combustion can occur. The nature of this mixing process is important. For example, the combustion of droplets is different when the liquid is completely vaporized prior to combustion than it is when individual diffusion flames envelop each drop. Differences such as these are considered in detail below.

Figure 59 illustrates two processes in which the droplets are completely vaporized prior to entering the combustion zone. The upper figure depicts a colloidal mixture of fuel and air entering from the left. The droplets evaporate as flow occurs from left to right, and are completely evaporated prior to entering the combustion zone. The resultant flame can be considered to be premixed. Thermal energy radiated from the combustion zone may accelerate the vaporization of the fuel as it approaches the flame. However, in many practical applications the fuel simply evaporates and in the process of so doing cools the resultant mixture. When this latter situation is the case, the rate of evaporation is influenced more by boundary layer considerations than anything else. Small drops present a larger area per unit mass of liquid from which evaporation can occur. However, this lower mass means that they are more likely to be accelerated to the velocity of the air stream in which they are enveloped. Thus the boundary layer through which the vaporized material must diffuse will be thicker than for a heavier drop, which is more likely to possess a larger relative velocity with respect to the surrounding air.

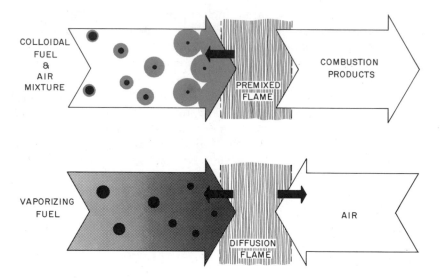

Figure 59. Combustion of liquid fuel droplets.

A somewhat different situation is shown in the lower portion of the figure. Like the previous situation, the fuel is completely vaporized prior to entering the combustion zone; however, there is no prior mixing of the fuel vapor with the air. In this case a diffusion flame is established where the two do meet.

A distinctly different situation exists when individual droplets are enveloped by diffusion flames as they simultaneously vaporize. This is illustrated in Figure 60. The upper portion of the figure shows a

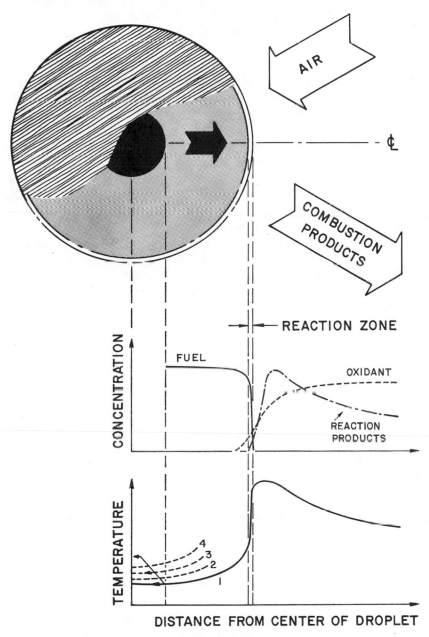

Figure 60. Combustion of a liquid fuel droplet enveloped in a diffusion flame.

droplet enveloped by a concentric spherical combustion zone. Vaporization of the droplet is driven by radiant energy supplied by the surrounding flame. The volume between the droplet surface and the inner bound of the concentric diffusion flame contains fuel vapor that is undergoing pyrolytic precombustion reactions as it flows outward to the combustion zone. Combustion occurs in a relatively thin zone where the relative proportions of air and fuel are favorable to sustaining a flame. Finally the combustion products are carried away by turbulent mixing.

The central portion of the figure shows the concentration of fuel, oxidant and reaction products as a function of radial distance from the droplet. The fuel concentration is essentially constant within the reaction envelope. It decreases slightly as it approaches the reaction envelope due primarily to the precombustion reactions and the presence of small amounts of air introduced by local turbulence in the flame zone. The concentration of oxidant decreases as the reaction zone is approached from the outside due to dilution of the air by the products of combustion. Finally the combustion or reaction products reach a maximum in the vicinity of the reaction zone and decrease in concentration as they are transported away from the combustion zone and diluted with the surrounding air.

The mixing that occurs between the combustion products and surrounding air is important. In the discussion of premixed flames the idea of two-stage combustion was introduced (refer to Figure 17). In the first or primary stage oxidation of the fuel produced partial and complete oxidation products such as H_2, CO, H_2O and CO_2. Subsequent oxidation of the partial combustion products was promoted by introduction of secondary air. In the case of droplet combustion an analogous situation exists. The thin zone surrounding the droplet can be thought of as the primary combustion zone. As the combustion products from this zone mix with the surrounding or secondary air any incomplete combustion products may be further oxidized. Additional information on droplet combustion will be found in References 2, 3, 4, and 5.

The concentration profile shown in the center portion of Figure 60 might be for a reaction product such as nitric oxide. The rate of formation of nitric oxide is a function of the oxygen concentration, the temperature and of the time the reacting gases are maintained at high temperature. As the reacting gases pass through the combustion zone, the rate of nitric oxide formation would be expected to reach a maximum since the temperature and the oxygen concentration increase.[6,7] After this the rate would drop to zero as the combustion products are quenched by the surrounding air. The concentration of carbon monoxide might be expected to be similar as discussed above. Some would be expected to form due to locally rich regions in the flame. Subsequent mixing of the combustion products with the surrounding air could either freeze the quantity of carbon monoxide already formed or promote its continued oxidation depending on the temperature of the secondary air.

The lower portion of Figure 60 depicts the temperature behavior of a multicomponent droplet and the precombustion gases surrounding it. At a given instant in time the temperature profile might be as shown by the solid line (denoted by the number one). The temperature within the droplet is constant while that of the fuel vapor increases as it approaches the vaporization zone. This can be thought of as due to the absorption of energy radiated from the flame by the particulate products of the pyrolytic precombustion reactions and also due to thermal conduction from the flame. As vaporization proceeds, the droplet will decrease in radius and its temperature will increase. The decrease in radius is denoted by the arrows and the corresponding increase in temperature by the dotted lines 2, 3 and 4. The increase in temperature is due to the progressive enrichment of the droplet in the heavier higher boiling fuel components as discussed earlier.

Whenever the liquid fuel contains relatively large amounts of heavy hydrocarbon the droplet behavior may differ from that described above. As the volatile components are heated, they diffuse to the surface of the drop. Some initial swelling or expansion of the drop may result if these volatiles are formed within the drop at a rate faster than they can diffuse through the heavier hydrocarbons. As volatilization continues, the temperature of the remaining liquid may rise to the point where liquid phase pyrolysis reactions occur. These reactions include cracking, dehydrogenation and polymerization and ultimately may lead to the formation of a viscous shell of asphaltines and resinous material. This "shell" further restricts volatilization. The net result is that during the process of volatilization the droplet may undergo several periods of swelling as volatiles expand and penetrate the viscous shell followed by contraction as the surface tension acts to reduce the interfacial area. Continued heating of the viscous unvaporized material can produce a vitreous mass that ultimately appears as a colorless glossy sphere in the effluent. The term *cenosphere* is used to describe such a combustion product.[8, 9]

Thus far two sources of particulate matter associated with the combustion of liquid droplets have been identified. First, there are particles that result from the vapor phase pyrolysis of fuel molecules as they travel from the droplet surface to the reaction envelope. If these particles are not completely oxidized as they pass through the reaction envelope, any further oxidation may be inhibited by their mixing with the cooler surrounding air. Second, there are those particles that result from the fusion of high molecular weight components before volatilization of the droplet is complete. These are termed cenospheres. A third source of particulates is ash- (inorganic) forming material, which is sometimes present in liquid fuels. Even when the hydrocarbons are completely vaporized this material will appear in the effluent as fine particulate matter, often an oxide, a sulfide or sulfate.

The concept of a spherical reaction envelope introduced in Figure 60 is obviously an idealization. Actual droplet combustion is considerably more complex. For example, when the burning droplet possesses motion relative to the air in which it is embedded, the reaction zone will not be concentric. Rather it will be distorted as shown in Figure 61. The distance of closest approach between the droplet and the re-

Figure 61. Wake flame enveloping a droplet moving relative to the surrounding air.

action zone occurs at the leading edge. Turbulence created in the wake of the droplet results in a tail as shown. If the relative velocity between the droplet and the air increases sufficiently, the wake will no longer be able to stabilize the flame. When this occurs the flame will blow off.

This of course changes the basic nature of the combustion process. First of all combustion, if it occurs at all, will take place somewhere downstream from the droplet. Second, any combustion that occurs downstream will be more like that shown in the top half of Figure 59 than like a diffusion flame. Third, the fuel vapor may well be distributed throughout such a large mass of air that it is well beyond the lean flammability limit and combustion will not occur at all. If the air temperatures are high enough, homogeneous vapor phase oxidation may occur. This can result in partial or even in complete oxidation of the fuel, depending upon the air temperature and the time available.

Combustion of Fuel Sprays

The fundamental phenomena associated with the combustion of liquid fuels were presented in the preceding section. The nature of the vaporization process is important because it is the primary determinant of whether combustion ultimately occurs as a premixed flame, a gaseous diffusion flame or as a group of diffusion flames each enveloping a droplet of liquid fuel. Liquid fuels are often prepared for combustion by dispersing them as a spray of droplets. This is done particularly when high combustion intensities (heat release per unit volume per unit time) are desired. It is this type of combustion, *i.e.*, burning of sprays, that is of primary interest in the present discussion.

In the preceding explanation of droplet combustion no consideration was given to the interactions that may occur between nearby droplets. Consider the combustion of many droplets in close proximity to one another. Figure 62 presents a two-dimensional representation of such

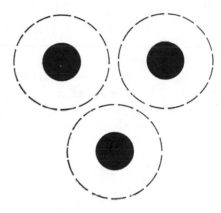

Figure 62. Combustion of a field of droplets.

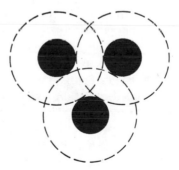

a field of droplets. For simplicity only three droplets are shown. One should imagine that these droplets are surrounded by others and the pattern is repeated in all three dimensions.

First consider the combustion of the three drops shown in the upper portion of Figure 62. The concentric dotted region surrounding each of the droplets depicts the volume of air required for complete combustion of that droplet. Since there is no overlap of these regions, combustion occurs under overall fuel-lean conditions. An alternative way of describing this situation is that on a mass-average basis combustion in the droplet field is fuel-lean. This is not the case in the lower part of the figure. The distance between adjacent droplets here is less and there is an overlap in the volumes of air that each would need for complete combustion. Obviously, in this case complete combustion is not possible and the available oxygen will be consumed before the fuel is completely oxidized, indeed, perhaps before all of the fuel is even vaporized. Combustion is rich on a mass average basis. In addition to the products normally associated with droplet combustion, relatively large amounts of partial oxidation products (such as CO) will also be produced.

In both of the cases shown in Figure 62 the energetics of combustion are markedly influenced by the close proximity of the droplets. Diffusion flames emit considerable radiant energy as discussed earlier. Combustion of a single isolated drop would be very nonadiabatic, because a considerable amount of energy would be radiated from the hot flame to its cooler surroundings. This is not the case when a large number of droplets in close proximity to one another burn. Each of the droplets radiates energy to its surroundings, but with the exception of those droplets near the edges of the droplet field, these surroundings are other burning droplets that are also radiating energy. Thus, since many burning droplets in close proximity merely exchange energy with one another, the combustion process will be more nearly adiabatic than might be first imagined. Factors such as this, which tend to reduce radiant energy transfer from the flame, also tend to increase flame temperature. This is particularly true in the case of Figure 62 (top) in which on a mass average basis combustion is fuel-lean. The availability of excess oxygen coupled with the increased flame temperature accruing from reduced radiation losses can be expected to increase the production of nitric oxide.[10] The same type of rationale cannot be applied to the situation presented in Figure 62 (bottom). In this case it is still true that the close proximity of many particles reduces the net radiant loss from each burning droplet. However, when the droplets are so closely spaced as to create fuel-rich combustion, not all of the available energy is released and the net flame temperature may well be considerably less than if the drops were to be widely separated during combustion.

It was noted earlier that combustion of a field of droplets under mass-average fuel-rich conditions may result in partial oxidation products; perhaps if conditions are rich enough unburned or even unvaporized fuel will remain after the flame has died out due to lack of oxygen. These products will be enveloped in the hot combustion gases generated by that portion of the fuel that did burn. Vaporization and pyrolytic reactions of these products will continue. Thus a combustion process such as this has the potential of producing large amounts of particulate matter and other pyrolysis products.

The field of droplets depicted in Figure 62 is shown as uniformly spaced. In any actual field of fuel droplets they will, of course, not be arrayed in such a uniform fashion. Rather, the distances between some of the droplets will be greater than that separating others. Furthermore these distances will be constantly changing as the droplets change position in both space and time. This constant change will be due to differences in the way different droplets, particularly ones varying in size, interact differently with the flowing air. This latter subject will be addressed in considerably more detail later on. For the moment it will be sufficient to recognize that variations, which are perhaps best described as statistical in nature, do occur. When the terms *rich* or *lean* on a mass-averaged basis are used to describe combustion in a field of droplets, it is necessary to remember that this refers to averages taken over a large number of droplets. Combustion of a given droplet or even of the droplets in a localized region may deviate considerably from the average conditions. Consequently, a field of burning droplets may be quite anisotropic with respect to combustion temperatures and the rates of pollutant formation. When this is the case, reduction in nitric oxide production will depend upon eliminating hot spots or other undesirable conditions within the burning droplet field. For example, reductions in particulate or smoke emissions may be accomplished by avoiding excessively rich conditions within some portion of the droplet field.

The three-dimension field of droplets that has been referred to might be produced in a variety of ways: for example, atomizing the liquid fuel with a nozzle, a spinning disc or ultrasonically. Figure 63 shows a number of techniques for fuel atomization.

Figure 63 *a* depicts a fuel spray that is created by forcing a liquid through an orifice (nozzle) by the application of hydrostatic pressure to the liquid. The work done on the fuel increases its internal energy. Part of this increase is due to the increase in kinetic energy associated with the high velocity droplets emerging from the orifice. Part is also associated with the work necessary to increase the liquid surface area when it is dispersed as a spray of fine droplets.

The subject of spray formation and behavior is very complex and only a very rudimentary treatment will be attempted here. There are,

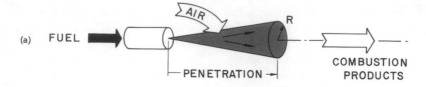

(a) FUEL

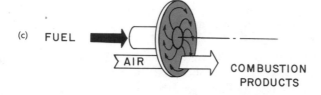

(b) FUEL

(c) FUEL

(d) AIR
    FUEL

(e) FUEL

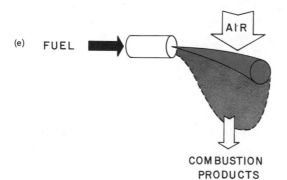

Figure 63. Atomization and combustion of a liquid fuel.

however, a number of topics of particular interest. These include the droplet size distribution within the spray, the geometry of the spray, and the way in which interactions between the droplets and the primary combustion air distort the geometry and distribution of fuel within the spray.

The size distribution of droplets in a spray is of interest for it influences a number of important factors such as fuel evaporation or vaporization rate and spray geometry. Many variables have an impact on the size distribution of fuel droplets obtained in a spray. One of these, the hydrostatic pressure applied to atomize the fuel, has already been identified. In general, when the pressure applied to the liquid increases, more work is done on the fuel and the droplet size distribution shifts in the direction of smaller diameters. Many physical properties of the fuel also influence the droplet size distribution. For example, as the viscosity of the fuel increases, the efficiency of atomization decreases and the droplet size distribution shifts in the direction of larger diameters. It is because of this that heavy hydrocarbon fuels are often preheated prior to atomization. The increase in fuel temperature decreases its viscosity, which in turn improves the atomization achieved. Other physical properties such as surface tension and density can also be of importance.

The relationship between droplet size distribution and spray combustion is complex. Consider what would take place if the droplet size distribution were to be reduced. The increased surface area per unit mass of fuel would increase the evaporation or vaporization rate of the fuel and thus allow more rapid mixing of the fuel vapor with the primary combustion air as shown in the upper branch of Figure 64. This in turn would increase the fraction of the fuel that burns in a premixed flame. The analysis is not complete however, for assuming that initial droplet velocity in the spray remains constant the mo-

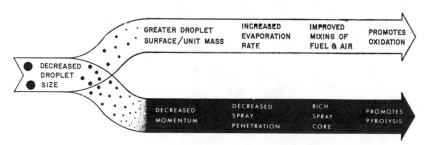

Figure 64. Droplet size, impact on the combustion process.

mentum of individual droplets is decreased as their diameter decreases. This will decrease the spray penetration and in effect reduce the volume in which the fuel undergoes primary combustion. In effect this decreases the air-fuel ratio in the core and tends to promote pyrolytic rather than oxidative reactions as shown in the lower branch of Figure 64. In any actual situations the types of changes outlined by the two branches of Figure 64 occur simultaneously. The net result in any given situation depends upon factors such as the initial starting point (mass average air-fuel ratio of the core) and the extent of the change (in average droplet size) that is made. In practice other factors, such as differences in the interactions between the air and particles of different sizes, also will affect the processes occurring in the spray core. This will be discussed in greater detail in a later section.

Figure 63 shows several alternative spray geometrics. Figure 63 *a* corresponds to injection of the fuel to form a conical spray with a small angle of divergence. The mass of fuel per unit mass of air within the cone is large. The extent of penetration of the cone is also shown. The only difference between 63 *a* and *b* is that in the latter the orifices are positioned to give a greater divergence of the individual spray trajectories. The average distance between fuel droplets will be greater for *b* than for *a*, assuming identical mass flow rates of fuel and of air. The greatest separation is achieved when the fuel is atomized by centrifugal force imparted to it by a rapidly spinning disc. This arrangement is illustrated in Figure 63 *c*.

Another method of atomizing liquid fuel is to compress a portion of the primary combustion air, combine this with the liquid fuel stream thus creating two-phase flow through the nozzle. This method, which is sometimes referred to as air-blast atomization, has the advantage of increasing the available oxygen in the case of the fuel spray relative to that which will exist in the cases discussed earlier. Figure 63 *b, c* and *d* all pertain to spray geometrics, which tend to increase the concentration of oxygen in the core of the spray.

The use of spray geometrics, which provide droplet separation analogous to that shown in Figure 63 *a* or which introduce oxygen into the spray core as per Figure 63 *d*, will promote more complete combustion of the fuel. From the point of view of emissions this can be expected to reduce particulate formation while at the same time increasing the amount of nitric oxide formed. A similar type of tradeoff was encountered earlier in the discussion of premixed flames. Selection of a rich air-fuel mixture for primary combustion minimizes the quantity of nitric oxide formed. However, this occurs at the expense of having to incorporate a secondary combustion step to oxidize the large quantities of carbon monoxide and other partial oxidation products formed during the primary combustion step. Furthermore, if it is desired to decrease the amount of nitric oxide even further this can be accomplished by introducing a reduction catalyst between the primary and secondary combustion steps. It is not generally possible to adopt the same approach to reduce emissions from combustion of a fuel spray. Relatively large excesses of air are usually required to insure complete combustion of a spray. This is true regardless of whether the mass-averaged condition within the spray itself is lean or rich. This excess oxygen is needed to insure complete oxidation of the pyrolytic pre-combustion reaction products and also to compensate for the local variations in mixture stoichiometry through the spray. During the droplet vaporization and combustion process (Figure 60), mixing of the combustion products with the secondary air surrounding the droplet occurs locally as well as concurrently. Separation of the primary stages (flame enveloping droplet and secondary mixing of combustion products with surrounding air) with the intent of providing a nitric oxide reduction step between them is not possible.

Catalytic reduction of the nitric oxide in the effluent from a combustion process with an overall lean mixture ratio requires the addition of enough secondary fuel to consume all of the excess oxygen and to provide net reducing conditions at the inlet to the reducing catalyst (Figure 30). This secondary fuel requirement imposes a more and more severe penalty as the amount of excess oxygen in the effluent increases. In many practical combustion devices, including those which have a net lean combustion zone to insure low concentrations of hydrocarbon and carbon monoxide when burning liquid fuels, it is a completely impractical approach. The amount of secondary fuel required may equal or exceed the fuel consumed in the primary combustion process.

From the above discussion it should be apparent that reducing the quantities of nitric oxide emitted from combustion of fuel sprays must be accomplished by preventing its formation and not by relying on secondary treatment. Limiting the nitric oxide formed in the spray zone requires very careful control of the conditions within the spray to prevent local hot spots and local regions of excess oxygen. This involves positive control over a number of factors such as the droplet size distribution, the angle of divergence of the spray cone, and the penetration of the fuel to mention but a few of the important variables. This is much more easily said than accomplished. Even when such control may be achieved under certain specific operating conditions, *i.e.*, flow rate of fuel and flow rate of air, it may be extremely difficult to maintain this control over the large variations of these flow rates that are encountered in practical combustion devices. It is also usually difficult to maintain this control during transient operation of combustion processes, *i.e.*, during the transition period from one steady state operating condition to another.

Some degree of reduction in nitric oxide formation during droplet combustion can be achieved by the use of previously discussed techniques, such as recycling of effluent gases and water injection. Both of these tend to reduce peak flame temperature and reduce the partial pressure of oxygen in the droplets' immediate environment. In the case of recycling effluent gases and substituting this recycled gas for a portion of the primary combustion air, the gains associated with reduction of oxygen partial pressure can be expected to be minimal when the oxygen content of the effluent gases is large. Injection of liquid water into the primary combustion air lowers the air temperature as it evaporates. This in turn will result in a lowering of the flame temperature, as was discussed in previous sections.

Interactions between Air and Spray

Thus far no consideration has been given to physical interactions between the fuel droplets in a spray and the air into which the fuel is injected. In actuality, complex interactions can and do occur. These interactions can alter the trajectories of the fuel droplets, stratify the droplets with respect to size within the spray, and even alter the nature of the combustion processes that take place.

Consider a conical spray as shown in Figure 63 *a*. If the air flow is perpendicular to the axis of the spray, the spray geometry will be distorted as illustrated in Figure 63 *e*. The interactions that occur are much more complex than just an alteration in the shape of the spray envelope. As the fine streams of liquid fuel emerge from the nozzle they break up to form droplets of different sizes. This process is illustrated in Figure 65. The different sized droplets behave differently in the flow field. The larger and heavier droplets possess the greatest momentum ($\frac{1}{2} MV^2$) and penetrate furthest into the air stream before their axial velocity component is dissipated by flow resistance. The trajectory of one of the largest droplets is depicted by path "A." Vaporization begins while the droplet is still in the spray core. The progressive vaporization also contributes to decreasing the droplet momentum as it penetrates further and further into the air stream. The point at which combustion actually begins depends upon a number of factors. Among them are the rates of vaporization and mixing of this vapor with the air, the temperature and pressure of the flowing air, and the presence of radiant energy from a nearby flame or other source of ignition to initiate and drive the precombustion reactions.

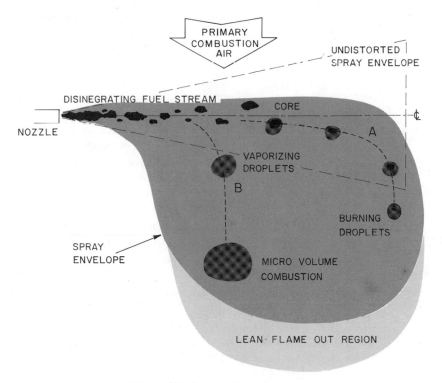

Figure 65. Combustion in a spray.

The smaller and lighter droplets possess much less momentum. They are rapidly accelerated to follow a streamline of the flowing air. This is depicted as path "B" in Figure 65. As they are carried downstream from the spray core they evaporate quickly due to their large surface area and may well evaporate completely before initiation of combustion of the vaporized fuel. The packets or microvolumes of fuel vapor will burn downstream in either diffusion or premixed type flames depending on the level of microscale turbulence in the flow field to facilitate mixing of the fuel vapors and air.

Each of the droplets in the spray–large, small, or intermediate–undergoes evaporation or vaporization from the instant it leaves the spray nozzle. Once a droplet is enveloped by a flame the fuel that is vaporized will burn in the vicinity of the droplet. However, prior to this the vaporized fuel will follow the air flow streamlines while the vaporizing droplet may have a different trajectory. This is particularly true for the larger droplets. This vapor, which then becomes separated from its droplet, is carried downstream. There it may either undergo microvolume combustion or, if mixing of the air and fuel vapor produces a mixture too lean to support combustion reaction, this portion of the fuel may be limited to partial oxidation or no oxidation at all. This region is referred to as the "lean flame-out region" (LFOR) in Figure 65 and may be a major source of any carbonyl and fuel species emitted from combustion of a liquid fuel as a spray.

The complexity of actual combustion of a spray should be apparent from the preceding discussion. It is seen that interactions between the air flow and the fuel droplets not only distort the shape of the combustion zone but can also result in different combustion regimes, *i.e.*, individual droplet, premixed and diffusion flame combustion, within the spray. The products and quantities of species emitted differ from one regime to another. These differences have been discussed separately in previous sections and will not be reiterated here.

Returning to Figure 63, this figure illustrates a few alternate geometries of fuel sprays and air flow. These were chosen for illustrative purposes. Obviously this figure does not exhaust the possible arrangements. From a very basic point of view some flow of air is obviously necessary to supply oxygen to the fuel spray and to remove the products of combustion. Now, the object of burning a liquid as a spray rather than in some larger aggregate such as a pool is to increase the combustion intensity (thermal energy released per unit volume per unit time). In order to burn a large quantity of fuel, a large quantity of air is also needed, more than can usually be supplied by natural convection. Consequently, forced convection of some type is generally employed to supply the air to burn a liquid fuel spray. This air may be introduced by flow in the same direction as the spray axis, or from some other direction, such as perpendicular to the spray axis as shown in Figure 63 *e*. The choice of the direction of air flow is made by the combustor designer and dictated by his objectives for a particular piece of equipment. In addition to the requirement that sufficient air be supplied to the primary combustion zone other factors influence the decision. For example, air flow may be chosen to minimize the possibility that incompletely vaporized fuel droplets will impinge on nearby combustor surfaces. Variations of the flows discussed above may include swirls at the periphery of the flame or films of air introduced at the surfaces to increase the thickness of the laminar boundary layer at the wall. This ordered flow and the turbulence associated with it is referred to as the primary flow or primary turbulence. The combustion process induces additional flow variations and turbulence as the hot combustion products expand within the confines of the combustor. This is referred to as secondary flow or turbulence. In some devices this secondary flow plays an important role in homogenizing combustion products, thereby promoting further oxidation of any partially oxidized or unoxidized species that may have survived the earlier combustion processes. These reactions, which can occur if the temperature remains high enough and if excess oxygen is present, are closely akin to those discussed earlier in the "Thermal Reactor" section.

Combustion of Liquid Pools

Relatively large aggregates or pools of liquid fuel (diameter or equivalent, 1 cm or greater) can be burned without first dispersing the liquid as droplets. The liquid must of course first be evaporated or vaporized and this vapor mixed with air to form a combustible mixture. Combustion of a pool of liquid is shown in Figure 66.

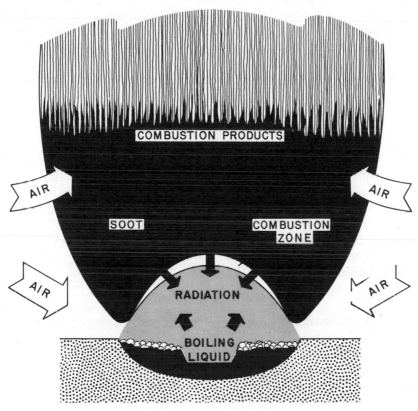

Figure 66. Combustion of a "pool" of liquid.

The processes that occur are not unlike the sequence discussed earlier in Figure 58. Perhaps the most unique aspect of pool combustion is that the large physical dimensions of the pool generally prevent adequate oxygen from reaching the interior parts of the combustion zone. The resultant flame tends to produce large quantities of particulate matter (sometimes referred to as smoke) and partial oxidation products, such as carbon monoxide and various carbonyls. Radiant energy transfer from such a flame is large. A portion of this energy is used to preheat and vaporize the fuel as well as to supply the necessary energy for the pyrolytic combustion reactions. The remainder is radiated to the cooler surroundings. This loss, however, is not usually counteracted by radiation from other nearby flames as was the case with spray combustion. The net result of this high loss of energy from the combustion zone is to lower the flame temperature. Mixing of the combustion products with additional air from the surroundings will reduce their temperature further. This quenching of the combustion products by cooler surrounding air often precludes any further oxidation of partial oxidation products and particulate matter. The quality of the combustion process can be improved somewhat by the use of forced convection to increase the supply of air to the combustion zone.

Figure 66 illustrates the situation in which a flame is stabilized above a pool of liquid. The major source of energy for vaporization is provided by radiant transfer from the combustion zone. A number of other variants are encountered in practice, for example, a film of liquid spread out over a surface constitutes a "pool." If this liquid film is not in direct communication with the flame, *i.e.*, it cannot "see" the flame or energy reflected by other surfaces from the flame, then other factors control the rate of evaporation or volatilization. For example, the temperature of the surface and the velocity of flow over the surface are important. The fuel vapor produced will be burned subsequently in a premixed or a diffusion flame as shown earlier in Figure 59.

Miscellaneous

Basically the combustion of liquid fuels occurs in one of two ways: a premixed flame or a diffusion flame. The diffusion flame may envelope the liquid or it may be far removed from the fuel. The nature and quantity of combustion products depends upon many factors as has already been discussed. There are some additional factors that warrant consideration.

The composition of a liquid fuel has an important bearing upon the combustion products. Some of the mechanisms by which the two are related have already been mentioned. For example, the presence of heavy hydrocarbon components in a multicomponent fuel can promote liquid phase pyrolysis reactions and ultimately lead to the formation of particulate matter. Some liquid fuels contain significant quantities of polynuclear aromatic compounds. These may survive combustion if they are trapped in the viscous mass that remains after the more volatile components of a fuel are vaporized. It should also be noted that regardless of whether PNA compounds are present in a liquid fuel, the conditions on the fuel side of a diffusion flame are conducive to their synthesis or synthesis of their precursors.

Liquid fuels may also contain important quantities of ash-forming and sulfur-containing compounds. When either or both of these species are present, secondary treatment processes other than those outlined in Figure 30 may be necessary to reduce the quantity of pollutant species present in the effluent. Indeed, the use of some of the techniques discussed earlier may be detrimental. For example, the use of an oxidation catalyst would be undesirable when the fuel contains significant quantities of ash-forming and sulfur-bearing compounds. The catalyst would promote the conversion of the undesirable sulfur dioxide to even less desirable sulfur trioxide. The ash on the other hand would tend to coat the catalyst and decrease its efficiency to promote oxidation. When fuels such as these are encountered, specialized noncombustion type treatments must be used.

4.4 COMBUSTION OF SOLID FUELS

Introduction

The range of solids burned in combustion processes is much more diverse with respect to fuel composition, and state of aggregation than is the case for gaseous and liquid fuels. For example, the most frequently encountered gaseous and liquid fuels are derivatives of natural gas or petroleum, and consist primarily of hydrocarbons. These include LNG, LPG, gasoline, fuel oils and kerosene. Occasionally other fuels such as methyl alcohol are encountered. A much wider gamut of solid fuels is encountered ranging from coal through wood, metallic substances to trash and explosives. Not only is the chemical composition of solid fuels quite variable, but large heterogeneities may be encountered within the same class of fuels. Trash for incineration is probably the best example of this. The state of aggregation of solid fuels may range from finely divided particles to objects as large as a building. This wide range of particle size adds further complication to the combustion of solid fuels.

Solid fuels contain a variety of chemical constitutents, not all of which are combustible. Thus, in Figure 67 the fuel is first subdivided

into combustible and noncombustible components. The latter are often
referred to as inerts. They are inert in the same sense the nitrogen of
air is inert: they may participate to some extent in reactions, particularly those at high temperature, but at the same time they do not contribute measurably to the energy released. Like nitrogen, they act as a
thermal sink and influence the peak temperatures achieved during combustion. Inert substances include the moisture that is absorbed by
many relatively porous solid fuels. A second type of inerts is the inorganic ash-forming substances. These include mineral substances such

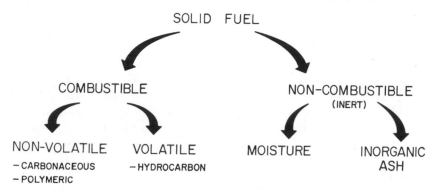

Figure 67. Chemical constituents that may be present in a solid fuel.

as silicates, sulfides, halogen salts, etc. When such substances are oxidized in the flame, the resulting ash is either retained in the combustion
bed itself or entrained in the effluent. In the latter case it is commonly
referred to as fly ash. The partitioning of this ash between the residual
material remaining in the combustion zone and entrained material
depends upon a variety of factors: weight and shape of the ash particle and the fluid velocity. The volatility of the oxide can also be important. The oxides of certain metals such as mercury, selenium and
cadmium have relatively high vapor pressures, and when they are
formed in the combustion zone a disproportionate fraction is present
in the postflame as a vapor. This vapor later condenses as the postflame gases cool and appears as ash, particulate, or a colloidal suspension in the effluent.

The combustible fraction of a solid fuel can be divided into volatile
and nonvolatile fractions. The nonvolatile portion is composed of
carbonaceous material, C_xH_y usually with $X \gg Y$ and relatively high
molecular weight. The volatile combustible fraction is comprised of
lower molecular weight compounds. The term *volatile* as used here

means those compounds that are volatilized at the temperatures to which they are exposed in the process of burning the solid fuel.

Organic species other than those containing only hydrogen and carbon may comprise a portion of either or both of the volatile and nonvolatile combustible fraction. For example, organically bound nitrogen, sulfur and halogens may be present. The presence of these species in the fuel is significant in so far as they may be converted to NO, SO_2, HCl, HF etc. during the process of combustion.

Fundamentals of Solid Fuel Combustion

The fundamental processes involved in the combustion of a solid fuel are shown in Figure 68. A solid fuel may or may not contain any volatile components. The illustration given is for the more complex case where volatiles are present.

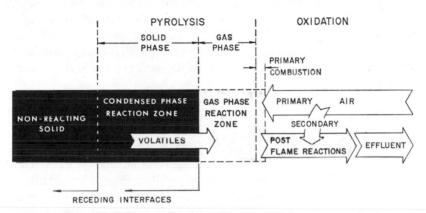

Figure 68. Combustion of a solid.

Basically in the case of a solid, reactions can occur in both the condensed and the vapor phases. The volatile components of the fuel, if any, are vaporized and flow away from the solid surface where they mix with the surrounding air. A diffusion flame is established within the envelope where the mixing of combustibles and oxidant forms a flammable mixture. This is denoted as the primary combustion zone.

As was true in the case of a liquid, strong thermal coupling exists between the diffusion flame, the vaporized fuel components and the condensed fuel phase. The region between the fuel surface and the diffusion flame contains little or no oxygen. Heating of the fuel vapors within this space promotes pyrolytic reactions of the same. This in turn promotes the nucleation and growth of particulate matter that emits a continuum of radiation as it passes through the high temperature combustion region. This radiation is the source of the yellow-orange flame coloration observed during the combustion of some solid fuels.

Like the particulate matter just referred to, the solid fuel surface also has high absorptivity and emissivity. The fuel surface may absorb a goodly portion of incident radiation. At the same time its thermal conductivity is relatively low, which restricts the rate of conductive dissipation of this energy to the interior portions of the fuel. A small portion of the energy is dissipated in vaporizing the more volatile fuel components and by endothermic pyrolytic reactions, which take place within the condensed phase. However, the total rate of energy dissipation by the processes of conduction, vaporization and chemical reaction is less than the rate of energy addition by radiation. Consequently the surface temperature rises until the solid fuel becomes incandescent and emits sufficient radiation energy itself to balance the radiant energy input from the flame. Of course, a portion of this reemitted energy is absorbed by the particles formed in the vapor space between the fuel and the flame, which promotes further pyrolysis. Pyrolysis is also likely as the volatile components pass through the pores of the high temperature surface region.

Solid fuels are particularly likely to produce particulate emissions. Some of these originate in the condensed and gas phase pyrolytic reactions discussed previously. Once these particles are formed their oxidation is a relatively slow process (refer to Chapter 2) and the time that it takes for these particles to pass through the primary combustion region may be insufficient to oxidize them. An additional source of particles is the solid fuel itself. High surface temperatures are achieved during combustion. Cracks and fissures of the surface material result. Particulate matter, in fact particles of rather large dimensions, may be dislodged in the process. Obviously, if the small particles nucleated by pyrolytic reactions are not consumed by the flame these larger ones will also survive and appear in the effluent.

Once the volatiles present near the surface of the fuel have been evaporated, oxidation of the residual solid (referred to as char when the original fuel is wood) occurs by a different mechanism. First, oxygen must diffuse into the surface. Then oxidation of the nonvolatile combustibles occurs. The energy liberated heats the surface, which in turn produces a characteristic glow. This process of "burnout" is a slow one since it is essentially a heterogeneous process involving several sequential steps, some of which are physical in nature (*e.g.*, diffusion).

The overall rate of combustion of a solid depends upon the individual rates of the processes already identified. These individual processes are evolution of volatiles, mixing of the fuel vapors and the oxidant, and the burnout of the nonvolatile combustibles. The rates of these individual processes depend upon the size of a fuel particle.

Thermal conduction within most solid fuels is a relatively slow process with respect to the rates of other processes encountered in combustion. The interior portions of a large fuel particle remain cool and only those volatiles near the surface are evolved. When a fuel particle has a large thermal thickness the interior portions remain virtually unchanged even as the surface region is undergoing degradative reactions. Only as the interfaces delineating the reaction zones recede (refer to Figure 68) are the interior volatiles heated.

One of the processes that can limit the rate of combustion of a solid is the availability of oxygen at or near its surface. In the case of small particles the available oxygen near the surface is increased by establishing a large relative velocity between the air stream and the particle itself. Either the air is forced to flow over a stationary particle or the particle is projected into the air stream. The relative motion between the two reduces the boundary layer thickness at the interface of the solid and air, thereby increasing the rates of oxygen diffusion and combustion product removal. In the case of burning large solid particles such as timbers or old box cars, it is obviously impractical to propel these objects through the air at a high velocity to reduce the boundary layer thickness and increase combustion rates.

As was the case for liquids where the particle size is small, diffusion of the oxygen and the combustion products occurs in spherical coordinates. The divergence associated with this case results in a steeper concentration gradient than is the case for a large particle where diffusion occurs in an orthagonal coordinate system. This increased concentration gradient associated with small particle combustion tends to increase the rate of combustion.

A final relationship between particle size and combustion rate revolves about the fact that combustion of a large solid particle involves a series of repeated steps. First the volatiles near the surface are burned and then the residual solid structure burns-out. As fresh unreacted solid is exposed, the process is repeated. The larger the solid particle, the greater the number of times this sequence must be repeated and the longer the time required for complete combustion of the solid.

Solid fuels may contain noncombustibles. The moisture is simply evaporated along with the combustible volatiles and carried away with the combustion products. A fraction of the energy produced by combustion must be expended to vaporize this moisture and heat it to flame temperature. It may also undergo dissociation as it passes through the flame. As a result the presence of moisture in the solid fuel reduces the flame temperature. Insofar as emissions are concerned this moisture tends to decrease the formation of nitrogen oxides for reasons previously discussed and to increase the particulate or smoke emitted.

In addition to moisture, solid fuels also often contain noncombustible ash-forming substances. These inorganic substances may be present as a variety of salts, sulfides, oxides, etc. They may increase their state of oxidation if high enough temperatures are achieved or they may simply be left when the volatile and nonvolatile combustibles are oxidized. Their ultimate fate in the combustion process depends upon a variety of factors. If the ash particles so formed are relatively small, they may be entrained by the flowing combustion products and appear in the exhaust as particulate matter. Since these particles are not combustible the thermal reactors or afterburners decribed earlier cannot be employed to reduce their concentration in the effluent. Special treatment devices such as electrostatic precipitators, fabric filters, cyclones, settling chambers and scrubbers must be used to effect their removal.

If high enough combustion temperatures are achieved, the ash may melt or fuse to form a slag. For example, silicates melt when the temperatures rise much above 1800°F. The molten globules may coalesce to form larger particles. On one hand this may be desirable because it is generally easier to remove larger particles from the effluent gases, but on the other hand the formation of slag may be undesirable also. It may be a solvent for the chamber materials. Also when the particles do coalesce, they may envelop other small particles of combustible material. If this occurs, the coated combustible material is generally lost to the combustion process and will not be consumed even if the slag particles are recycled to the combustion zone. Thus the overall combustion efficiency achievable (Equation 34) is decreased because not all of the chemical energy in the fuel can be released. Another important consideration is that the molten slag may adhere to and corrode the internal surfaces of some combustors.

In addition to the substances considered above there are a number of additional ones that are sometimes present in solid fuels. Sulfur may be present in both the inorganic and organic components of the fuel. Regardless of its chemical form this sulfur is oxidized in the flame to sulfur dioxide (and as discussed earlier to small amounts of sulfur trioxide). Halogens (present for example in many synthetic polymeric solids) can be converted to their corresponding acid anhydrides or to oxygen-containing species. For example, chlorine may appear in the effluent as HCl.

Solids may also contain organic nitrogen compounds, for example, pyridine. This presents some peculiar problems with respect to the control of nitric oxide. At high temperatures, nitric oxide is formed by the fixing of nitrogen present in the air. The addition of a small amount of fuel nitrogen would not materially alter the high temperature equilibrium concentration of nitric oxide in the flame. However, because the amount of nitric oxide formed at high temperatures is generally unacceptable regardless of whether the fuel contains nitrogen, various control measures that reduce flame temperature are often employed. As the flame temperature decreases, reaction rates do likewise. As reaction rates decrease, the chemical composition of the reacting gases deviate further and further from equilibrium. Oxidation proceeds more and more as a series of elementary steps not unlike those discussed in the precombustion section, Equations (14) and (16). Fuel nitrogen is thus oxidized to nitric oxide, which is not subsequently decomposed by passage through a high temperature zone where the species present are equilibrated.[11,12] In dealing with solid fuels containing nitrogenous compounds, one may notice that nitric oxide emissions may exhibit a minimum behavior as progressive control measures are taken. At high temperatures large amounts of nitric oxide are formed due to the fixation of nitrogen in the combustion air. As control measures are taken to reduce flame temperature, nitric oxide formation first decreases and then increases again due to the contribution of fuel nitrogen at low combustion temperatures.

Combustion in Fuel Beds

For a variety of reasons, many of which were discussed in the preceding section, solids are often subdivided into relatively small particles and burned in a "bed." Such an arrangement increases the rate of combustion of the solid and reduces radiation losses. The energy radiated by incandescent particle surfaces is large, so large in fact that if the solid is not "facing" another burning solid, the surface temperature may drop to the point where combustion ceases. This is best illustrated by an ordinary campfire. An isolated log does not burn easily by itself. However, when a group of logs are placed close to one another, combustion readily occurs on the interior surfaces that face one another. The radiant losses from one log are balanced by a gain in radiant energy from nearby ones. Thus, the use of a bed of close-packed solids is not just one of convenience or of increased reaction rates; it is essential to maintain combustion. A variety of possible bed arrangements exists with regard to fuel and air supply, some of which will be discussed below.

Overfeed

One arrangement, called *overfeed* because fresh fuel is introduced at the top of the bed, is illustrated in Figure 69. The air supply is

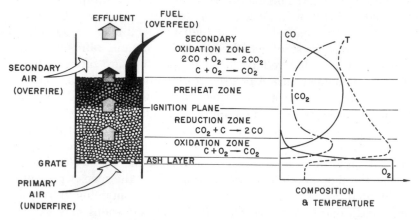

Figure 69. Combustion of a bed of solid fuel, overfeed arrangement.

divided between the primary combustion air, which is introduced at the bottom of the bed, and the secondary air, introduced above the bed. This is analogous to the arrangement of processes shown in Figure 17. The amount of primary combustion air controls the rate of combustion in the fuel bed. This follows from the fact that the fuel cannot be consumed at a rate any greater than the available oxygen permits. Furthermore either a deficiency or an excess of air will reduce the temperature and the rate of combustion. Control of the quantity of primary air is one means of regulating the bed temperature to prevent slag formation. The amount of secondary air, on the other hand, controls the overall combustion efficiency. Sufficient secondary air must be added to completely oxidize any unburned or partially oxidized species originating in the bed.

Fresh fuel is supplied to the top of the bed as shown. If this fuel contains any volatile components, these will be vaporized as the fuel is preheated by the hot gases flowing through the fuel. The significance of this, *i.e.*, preheating the fuel with hot combustion products, will be discussed in some detail later on. Actually for large fuel particles volatile components deep in the interior may not all be vaporized by the time the fuel arrives in the combustion zone. For the moment, however, assume that by the time the fuel reaches the ignition plane all of the volatiles have been removed and combustion can be thought of as that of pure carbon. The primary air is preheated by the ash layer and first contacts the fuel at the bottom of the oxidation zone. Here the oxygen and fuel combine

$$C + O_2 \rightarrow CO_2 + 94 \ \frac{kcal}{mol} \qquad (17)$$

to produce carbon dioxide, a product of complete oxidation. The heat of reaction (combustion) results in a large rapid rise in the bed temperature. This reaction continues as the gases rise through the bed of fuel until the oxygen contained therein is depleted.

The hot oxygen-depleted combustion products continue to rise through the bed. The carbon dioxide and water vapor present in these gases reduce the unburned fuel above.

$$CO_2 + C \rightarrow 2CO \qquad (79)$$
$$H_2O + C \rightarrow H_2 + CO \qquad (80)$$

These very endothermic reactions occur only at high temperatures. The temperature of the gases rising through the bed decreases both as a result of these reactions and any heat transfer from the bed itself. Note also that the concentration of carbon monoxide, a partial oxidation product, increases at the expense of the carbon dioxide formed in the oxidation zone. Figure 69 shows only carbon monoxide and carbon dioxide. Other species such as hydrogen, methane and other low molecular weight hydrocarbons are also present.

The gas entering the preheat zone is hot and contains reducing species, i.e., carbon monoxide, hydrogen, etc. This gas preheats the fuel and vaporizes any moisture or volatile organic species that are present. Since little or no oxygen is present, these volatile organics can only undergo pyrolytic reactions to produce particles and other products characteristic of pyrolysis. The smoke produced may be shades of white, yellow or black depending upon the fuel composition. Secondary air is added above the bed to promote the oxidation of particles and other partial oxidation and pyrolysis products. In essence the gas volume above the bed is used as a thermal reactor.

A number of factors are important with regard to the overall combustion efficiency achieved. The quantity of secondary air and its temperature influence the ultimate gas phase reaction temperature. Care must be taken not to quench the reactions in this region and thereby produce carbonyls or other additional partial oxidation products. The point at which the secondary air is added and the method of introduction influence the effectiveness with which the secondary air and combustion products are mixed. Usually an excess of 30 to

50% air (overall excess, *i.e.*, total of the primary and secondary air relative to the primary fuel) is required to allow for incomplete mixing.

Finally the emissivity of the wall surrounding this gas volume can have an important bearing on the extent to which particles are oxidized. Particles emitted by the bed may arise in a variety of ways. One route, formation by pyrolytic reaction of gases in the reduction and preheat zones has already been addressed. A second possibility arises because as the fuel particles react, their mass and diameter decrease. A point may be reached at which the smaller particles are entrained by the combustion gas before they are completely consumed. This is particularly true when channeling of gases through the fuel bed results in localized areas of high gas velocity. Those particles that are not completely oxidized above the bed and that can be separated from the effluent are recycled to the bottom of the bed for further reaction. This, of course, does not apply to particles composed of unburned fuel and fused ash. Neither thermal reactions above the bed nor recycling through the bed will be effective in recovering this fuel.

Combustion of solid fuel in a bed can be either a steady state or a batch process. For example, any ash that remains in the bed shown in Figure 69 accumulates on the grate at the bottom. If arrangements are made to continuously remove this ash, a steady state can be achieved. If the ash merely accumulates, then the combustion process will be a batch one and must be periodically shut down to remove the accumulated ash. Likewise the feed may be steady state or batch. The batch case is of greatest interest here because of its impact on emissions. When the fuel is dumped on top of the burning bed in batch fashion, a number of processes are upset. The temperature and composition of the gases leaving the bed change as the fresh fuel is preheated and the volatiles contained therein are evaporated. This in turn alters the secondary combustion reactions above the bed. The effluent gas will contain its greatest load of combustible species during that time interval when its temperature is low. This tends to reduce the overall combustion efficiency.

One final observation with respect to Figure 69. The combustion zone is shown as a thin stationary zone within the bed. This implies that the ignition and burning rates of the fuel are approximately equal. This may or may not be the case. Ignition is a complex phenomenon. The rate of ignition depends upon many factors including the homogeneity of the fuel with respect to particle size, chemical composition, and moisture content. If the ignition rate is greater than the burning rate (more likely for large particles with low moisure content), most of the fuel will be consumed after the ignition front has reached the top of the bed. Conversely, if the ignition rate is less, the burning rate is unstable and combustion will cease.

Underfeed

An alternate arrangement for solid fuel combustion is shown in Figure 70. Here the fuel is introduced at the bottom rather than the

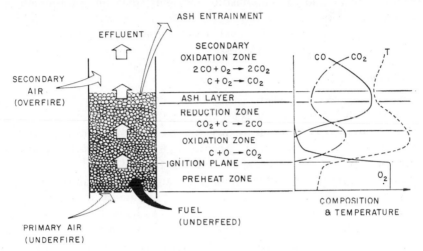

Figure 70. Combustion of a bed of solid fuel, underfeed arrangement.

top of the bed. The flow of primary combustion air and fuel is co-current rather than countercurrent. With this arrangement the fresh fuel is not preheated by the hot combustion products. Fuel preheating, if it occurs at all, must be accomplished by heat transfer from the primary air. Evaporation of volatiles and oxidation of the residual solid fuel occur in the same region within the bed. Consequently, underfeed is conducive to promoting the oxidation of the volatile organics rather than pyrolysis as occurs when fresh fuel is introduced on top of the bed. Moisture present in the fuel is also released in the oxidation zone. The energy absorbed in the process reduces the adiabatic flame temperature.

The combustion zone for underfeed is subdivided into an oxidation and a reduction zone as was true for the overfeed arrangement. Transition occurs from oxidation to reduction when the oxygen in the primary air is consumed. The principle reactions in these zones are similar for the two cases. As the fresh fuel is introduced at the bottom, the ash is carried to the top of the bed. Here it can be removed either by entrainment in the combustion products or by other physical means.

Secondary air is added above the bed to complete the oxidation of partial combustion products generated in the oxidation and reduction zone. For reasons just discussed underfeed is less likely to produce gaseous organic species that must be oxidized above the bed than is overfeed. At the same time, since the fuel particles are progressively reduced in mass and size as they flow cocurrently with the primary combustion air and the combustion products, particle entrainment is more likely to occur. This requires care to insure that these particles are oxidized above the bed, recycled to the feed or, in the case of ash, separated from the effluent for disposal.

Fluidized Beds

As the primary combustion air flows upward through the fuel bed it exerts a force on the particles in its path. As the velocity of the flowing air increases, this force also increases. If on the average the force exerted on the particles is great enough to overcome the opposing force of gravity, the bed becomes fluidized. In other words, the flowing primary combustion air rather than the grate supports the bed of particles. The particles within the bed are not stationary with respect to one another, but rather circulate in a random fashion. This arrangement has several aspects that present both unique advantages and problems when the bed undergoes combustion. This arrangement is shown in Figure 71.

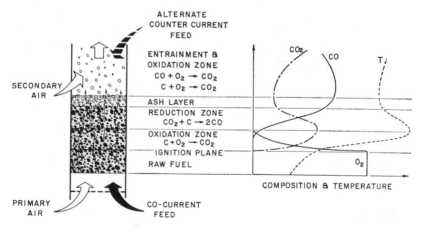

Figure 71. Combustion of a bed of solid fuel, fluidized bed arrangement.

The primary and secondary air are arranged as they were previously. The primary combustion air must support the bed. Therefore its velocity as it flows through the bed of particles is critical. If the velocity is too low, the bed will settle and combustion will degenerate to one of the two cases previously discussed (depending upon the feed location). If the velocity is too great, an inordinate amount of unburned fuel will be entrained by the flowing gases. The question of correct velocity is not as simple as it may appear from the preceding statement. Even the raw fuel is not generally nondispersed with respect to particle size and mass. Furthermore, as the particles are oxidized their mass and size continually decrease. The velocity that is correct for supporting one size particle will be too great or too small for other particles.

The feed may be introduced from the bottom as shown, as well as from the top. The greater mobility of particles within the bed will tend to cause the smaller lighter particles to migrate to the top and to be entrained in the effluent. Ideally the fuel and the ash could be separated. The fuel particles could then be recycled to the combustion bed while the ash is disposed of.

Specific zones, raw fuel, oxidation, reduction and ash are shown. These are introduced only for the purpose of indicating the composition and reactions in a general way. The random particle motion results in much more diffuse boundaries than was true for combustion in stationary beds.

As a result of the intimate contacting of the primary combustion air as it follows a tortuous path through the bed of moving particles, less excess air is necessary than for the previous cases of a stationary combustion bed. For example, an overall excess (primary plus secondary air) of 5% may produce the same overall combustion efficiency as is obtained with 30% excess air in a stationary combustion bed. The decreased quantity of excess air is advantageous inasmuch as the volume of the combustion and ancillary equipment, air handling and effluent treatment can be substantially decreased. This, of course, represents a potential equipment cost savings. At the same time a decrease in the amount of excess air results in higher bed temperatures, which in turn can increase nitric oxide and slag formation. To minimize these problems some fluidized combustors are equipped with internal heat transfer surfaces to limit the bed temperatures. In some applications, such as incineration, a portion of this energy that would ordinarily be lost in the hot combustion products can be recovered by using it to generate steam for space heating or for conversion into mechanical energy by means of a turbine.

In addition to the incorporation of steam-generating coils in the combustion bed, a number of other techniques have been or are being developed for use with fluidized bed combustors. Operation at elevated pressures can further reduce the size of the equipment, particularly the gas-handling equipment used to treat the effluent gases. Inert material such as alumina or silica particles can be added to the bed to maintain more uniform temperature throughout the bed. This is particularly desirable when the bed is being used for steam-generating purposes. The thermal inertia of the inert particles prevents rapid fluctuations in bed temperature, which occur when the feed rate or composition undergoes temporary excursions. Less inert material such as calcium oxide particles can also be added to the bed. Such materials will react with at least a portion of any sulfur in the fuel to produce sulfate particles and reduce the quantity of gaseous sulfur dioxide produced by combustion.

Piles and Heaps

The preceding discussion of combustion in overfeed, underfeed and fluidized beds presupposes a degree of control over the process, that is of the rates and distribution of primary and secondary combustion air. In many cases solids simply burn in a pile of some sort. Ignition may be intentional or unintentional. An example of the latter is spontaneous ignition caused by exothermic chemical reactions deep in the pile, which raise the temperature of the material to its ignition point. Whichever is the case, it is generally the lack of control over basic combustion variables such as the air-fuel ratio that has the greatest impact on the emissions of a burning or smoldering pile of solids.

This case is probably best characterized as one in which there is poor air distribution with respect to the fuel. Air is drawn toward the pile by the draft created when the hot combustion products rise (see Figure 72). Generally this air does not penetrate deep into the pile. The

Figure 72. Combustion of a bed of solid fuel, pile arrangement.

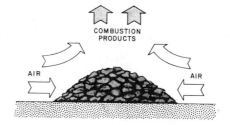

material near the surface burns, while that deep within the pile either smolders if it is heated sufficiently or simply does not react at all. The smoldering process is one in which the fuel does not burn as such due to the very rich local air-fuel ratio. At the same time, however, the volatile species are evolved and undergo pyrolytic reactions due to the elevated temperatures to which they are exposed as they mix with the hot combustion products from other parts of the pile. The result is that relatively large quantities of partial oxidation and particulate matter are produced. These products are subsequently quenched when they mix in an uncontrolled fashion with secondary combustion air above the pile. Consequently burning of solids in a pile usually results in copious smoke and other undesirable combustion products. Some degree of control can be achieved by the use of a forced draft to improve air penetration in the pile. However, the results are generally less desirable than can be achieved by the previously discussed techniques for burning solid fuels.

4.5 PROPELLANTS

Propellants are those solids containing both fuel and oxidant (the oxidant here may be something other than oxygen). It may consist of a single compound that decomposes to produce species that subsequently react to produce a large quantity of gaseous reaction products. Or, propellants may be composed of a physical mixture of solids that react with one another to produce products that are largely gaseous.

Examples of propellants are explosives and solid rocket propellants. Chemically speaking, a wide variety of substances may function as propellants. These substances may contain carbon, hydrogen, oxygen, nitrogen, sulfur, halogens, metallic elements, etc. The oxidation or reduction products of these elements will of course be present in the combustion products. Therefore it is necessary to examine the chemical composition of each particular propellant to gain insight into the possible nature of the species that will be emitted.

The processes involved in propellant combustion are shown in Figure 73. The rate of reaction (conversion from solid to gaseous reaction

Figure 73. Combustion of a propellant.

products) depends upon a number of factors. One, of course, is the chemical composition. Another is the exposed surface area at which the reaction occurs, and a third is the physical state of the propellant. For example, a propellant that is loosely packed will burn more rapidly than if it were densely packed. Finally, solid fuels of this nature sometimes undergo detonation reactions. Detonation is a very complex subject and beyond the scope of this discussion. However it should be noted that when detonation does occur the gas/solid interface propagates very rapidly through the solid (at the speed of sound in the solid). Very high temperatures are achieved as this interface, or shock wave, converts the solid to gaseous reaction products.

As pointed out earlier the chemical composition of the reaction products reflect the law of conservation of mass. If an element is present in the propellant, it will be present in the reaction products. In addition to this simple observation, it is worth noting that due to the very rapid and even violent nature of propellant combustion large quantities of particulate matter, ranging from unburned fuel particles dislodged from the surface during combustion to parts of the container material, may be present in the effluent.

Another factor that influences the combustion products of a propellant is the high temperatures often achieved during combustion. This is followed by rapid expansion and therefore thermal quenching of the reaction products. The high temperatures result because both fuel and oxidant are present (a coolant of some sort may be blended with the solid but this is not necessary). Without the thermal inertia of the diluent, higher temperatures and a greater degree of dissociation into elementary and simple diatomic or triatomic species will be achieved. The rapid expansion that usually follows will tend to preclude recombination reactions forming more complex products. Thus the reaction products will tend to reflect the species present at peak flame temperatures and pressures. This, in fact, is generally desired, for limiting the reaction products to simple diatomic and triatomic species produces a larger volume of effluent gases than would be the case if more complex higher molecular weight species were produced.

4.6 SUMMARY

A diffusion flame is a more complex process than a premixed one. Physical processes such as evaporation of liquid fuel or volatilization of solid fuel and, in either case, mixing of the resulting vapor with the oxidant may control the overall reaction rate. This is in contrast to a premixed flame where chemical factors are rate determining. Nonuniformity of the size of the liquid droplet or solid particle further complicates the situation. Pyrolysis in the precombustion region may produce species that resist oxidation in the combustion zone. Combustion does not take place at a particular air-fuel ratio but over a range of ratios. These are just a few of the complicating factors that must be considered in an analysis of diffusion flames.

In principle, models of diffusion flames can be constructed by developing submodels for each of the individual important processes and by then combining these submodels in a way to describe adequately the overall process. In practice, the complexities, the heterogeneities, and the statistical nature of a diffusion flame combine to make this rather impractical. Consequently, whereas numerous thermodynamic and kinetic models of premixed flames exist, few are available for diffusion flames. This is particularly true for cases such as droplets enveloped by diffusion flames where wakes and other phenomena are involved. Most of the information on the formation and emission of trace species from diffusion flames has been obtained experimentally. Insofar as the emissions themselves are concerned this does not present any insurmountable difficulties, for measurements can be made on just about any source. Severe problems are encountered, however, when attempting to make experimental measurements to elucidate the mechanisms and intermediate species that contribute to the formation of the emitted species. Geometric factors in a particular situation influence the radiant transport of energy between the different zones in a flame. This important feedback mechanism couples almost all of the physical and chemical processes that occur. Factors such as this make it extremely difficult to construct the type of laboratory experiments that are suitable for mechanistic studies and that are at the same time adequate representations of the process being modeled. Thus, it is this complexity that not only defies mathematical modeling but experimental studies as well, making the understanding of diffusion flames one of the most challenging problems in the field of combustion.

REFERENCES

1. Eligo, R., N. Nishiwaki, and M. Hirata. "A Study on the Radiation of Luminous Flames," *Proc. Eleventh Symp. (International) on Combustion* (Pittsburgh, Pa.: Combustion Institute, 1967) p 381.
2. Udelson, D. G. "Geometrical Considerations in the Burning of Liquid Drops," *Combust. Flame,* 6, 93 (1962).
3. Hottel, H. G., G. C. Williams, and H. C. Simpson. "Combustion of Droplets of Heavy Liquid Fuels," *Proc. Fifth Symp. (International) on Combustion* (New York: Reinhold Publishing Co., 1955) p 101.
4. Spaulding, D. B. "The Combustion of Liquid Fuels," *Proc. Fourth Symp. (International) on Combustion* (Baltimore: The Williams & Wilkins Co., 1953) p 847.
5. Longwell, J. P. "Combustion of Liquid Fuels," in *Combustion Processes, Vol. II, High Speed Aerodynamics and Jet Propulsion Series,* B. Lewis, R. N. Pease, and H. S. Taylor, Eds. (Princeton, N.J.: Princeton University Press, 1956).
6. Tuteja, A. D., and H. K. Newhall. "A Study of Nitric Oxide Formation in Diffusion Flames," presented at the 1973 Technical Session of the Central States Section of the Combustion Institute, Champaign, Ill. (March 1973).
7. Tuteja, A. D., and H. K. Newhall. "Nitric Oxide Formation in Laminar Diffusion Flames," in *Emissions from Continuous Combustion Systems,* W. Cornelius, and W. G. Agnew, Eds. (New York: Plenum Press, 1972) p 109.

8. Kobayasi, K. "An Experimental Study on the Combustion of a Fuel Droplet," *Proc. Fifth Symp. (International) on Combustion* (New York: Reinhold Publishing Co., 1955) p 101.

9. McCrone, W. C., R. G. Draftz, and J. C. Delly. *The Particle Atlas* (Ann Arbor, Michigan: Ann Arbor Science Publishers, 1967) p 110.

10. Henein, N. A. "Combustion and Emission Formation in Fuel Sprays Injected into Swirling Air," presented at the SAE Automotive Engineering Congress, Detroit, Mich., paper no. 710220. (January 1971).

11. Breen, B. P., A. W. Bell, N. B. DeVolo, F. A. Bagwell, and K. Rosenthal. "Combustion Control for Elimination of Nitric Oxide Emissions from Fossil-Fuel Power Plants," *Proc. Thirteenth Symp. (International) on Combustion* (Pittsburgh, Pa.: The Combustion Institute, 1971) p 391.

12. Bartok, W., A. R. Crawford, and A. Skopp. "Control of NO_x Emissions from Stationary Sources," *Chem. Eng. Progr.*, **67**, 64 (1971).

APPLICATIONS

The discussions of fundamentals and applications have been separated in this book, with only occasional references to applications made in Chapters 1 through 4. This chapter will provide the reader with a guide to sources of information on some of the more important combustion-related sources of air pollution.

No attempt is made to provide an exhaustive review of the literature; rather, the references cited here were chosen to provide a picture of recent developments. By consulting these references, the reader will in turn find many other references for specific details and historical developments. It is important to realize that only recently has the emission of air pollutants become a serious constraint for the design and operation of combustion processes. Consequently, many of the most significant engineering advances have been developed in the past few years and will continue to develop in the years ahead. Much of the literature cited will be quickly dated. To remain at the forefront, one must consult the latest issues of journals and participate in professional society meetings where the results of current research are being presented. Some of the latest developments are proprietary in nature and as such will not be in the public domain. Personal dialogue with those who are doing the work is without doubt one of the best ways to remain current on what is available.

There are a number of sources of information on emissions from combustion-related sources that the reader should become familiar with. These include journals, professional societies, abstracting services and government-sponsored reports and publications.

Air pollution is a subject of contemporary concern, as was noted in Chapter 1. As a result, a large number of publications ranging from trade magazines to scientific journals contain articles dealing with source emissions. Some are strictly descriptive, others are highly mathematical and technical in nature. A few sources that the reader may find to be particularly helpful for articles on air pollutants emitted by combustion processes follow. Professional organizations that conduct technical meetings and publish journals include: Air Pollution Control Association,[1] The Combustion Institute,[2] American Petroleum Institute,[3] American Society of Mechanical Engineers,[4] Institution of Mechanical Engineers,[5] American Chemical Society,[6] American Institute of Chemical Engineers,[7] and the Society of Automotive Engineers.[8] A number of international journals will also be found to contain relevant information.[9-12]

As a result of the Clean Air Act and its subsequent amendments[13] agencies of the U.S. Government were directed to undertake studies on air pollution. The U.S. Department of Health, Education and Welfare (HEW) was responsible for this prior to November, 1970, when the U.S. Environmental Protection Agency (EPA) was created. Reports of these agencies are published in two series identified with the prefixes AP[14] and $APTD$.[15] An index of these reports is available from EPA.[16] Various other government agencies are also engaged in studies relating to the emission of trace species from combustion sources.[17,18]

In addition to these sources, agencies within the United Nations have conducted studies and issued reports that may be of interest.[19] Finally, it should be emphasized that one of the basic problems that must be tackled when dealing with the emission of trace species is that of how to measure them. Standardized methods are needed so that the measurements made by different investigators, at different locations, and at different times can be compared. The American Society for Testing Materials[20] is one organization that publishes such standards.

The approach adopted in this chapter is to provide the reader with a narrative that places the literature cited in perspective. It is hoped that this will be of greater value than the alternative of simply listing the literature with no explanation.

5.1 STATIONARY SOURCES

Most combustion processes can be categorized as either stationary or mobile. This classification scheme was used in Table I. For purposes of discussion here, stationary sources are divided into three subclasses. The first is boilers, which are used to generate steam (and to a lesser extent hot air) that in turn is used to generate electric power and to heat buildings. The second subclass is the incineration of waste materials (gaseous, liquid and solid wastes). Finally, there is industrial processing.

The evolution of approaches to the control of pollutants from stationary process is presented in Figures 74, 75 and 76. These figures are conceptual and can be applied to many diverse sources of air pollutants, those that are combustion related as well as those that are not. A combustion process may be thought of as a special case. In

Figure 74, the air is the source of oxidant and the fuel is the raw material. With the exception of applications where water is injected to control nitric oxide, water is not normally a reactant for combustion. Any unreacted air will be exhausted along with the products of combustion that may include solid, liquid and gaseous species as shown in Figure 74.

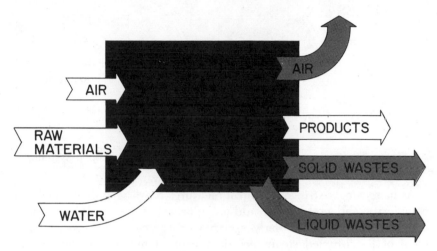

Figure 74. Flow streams for a process.

The most elementary form of control is to adjust the controllable variables to reduce the quantities or types of species emitted. An example of a variable that can be controlled in some processes is the equivalence ratio of the primary combustion process.

An alternate and often complementary strategy is the use of add-on devices. This approach is illustrated in Figure 75. An add-on device is placed in the effluent stream to remove trace gases and suspended particles. A detailed description of these devices is beyond the scope of this discussion. However, since they do find widespread application in the treatment of effluents from stationary sources, some literature references are provided here.

Filters,[21-25] cyclone separators,[21,23,26,27] and electrostatic precipitators[21,23,28-30] are used to remove particulate matter. Scrubbers or absorbers[21,23,26,31-34] and adsorbers[21,22,35] are used to remove trace gases. Aerosol substances can be removed by mist eliminators.[36,37,38]

General discussions of control techniques that traditionally have been used to reduce the emissions from stationary sources can be found in References 39, 40, 41.

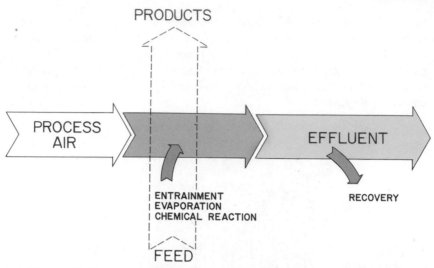

Figure 75. Recovery of trace emissions from a process, open system.

As increasingly stringent control measures have been demanded, the complexity and cost of the add-on approach to control has rapidly increased. This is particularly apparent in the case of stack-gas treatment of fossil-fueled steam-electric power plants. Large amounts of material removed from the effluent cannot be disposed of as liquid or solid wastes because of simultaneous pressures to reduce the emissions of these materials too. What began as an add-on has evolved into a complex chemical plant with all of the attendant problems thereof.

The pressures identified above are leading to the development of closed-system technologies, which are sometimes referred to as alternate nonpolluting technologies. The concept of a closed-system process is illustrated in Figure 76. Rather than the use of air as a working fluid in the process, an alternate approach is chosen. The working fluid is treated to remove contaminants before it is recirculated. The substitution of a nuclear for a fossil-fueled steam-electric plant is an example of such an approach. Care must be exercised in the use of the term *alternate nonpolluting source* for in some cases the problem is merely an exchange of one form of pollution for another, or a shifting of emissions from one geographic location to another.

Stationary Boilers

Stationary boilers are generally used to produce steam, and they find widespread application in the production of electric power, the heating

PRODUCTS

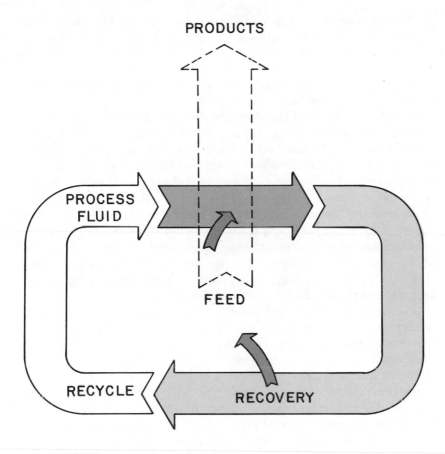

Figure 76. Recovery of trace emissions from a process.

of buildings (space heating) and industrial processing. For the most part, the energy used to produce the steam is derived from the combustion of fossil fuels. Texts are available dealing with the technology of boilers and applications thereof.[42, 43] The focus of these texts is on subjects other than the emissions produced by the combustion process. Literature reviews have been published that do address the subject of emissions from this class of stationary sources.[44, 45]

One useful way of classifying boilers is by size. The largest are those used by utilities for electric power generation. Most of these units have an electrical generating capacity of 100 megawatts (MW) or greater. The trend is toward larger and larger units, and several of the units being built will exceed 1000/MW while future ones may have a capacity of 1500/MW.

Most commercial and industrial processes use smaller units. These intermediate-sized boilers are used to supply process steam as well as to generate on-site electrical power. At the other end of the size spectrum are the small boilers used for space heating of residences. In some cases, steam is produced while in others the combustion process merely heats air or water.

Another useful way of classifying boilers is by the fuel burned. In the case of fossil-fueled boilers, this is generally either natural gas, petroleum or coal. There has been a trend over the past years to switch from coal to oil. There are two main reasons for this: economic and environmental. In the case of the large industrial and utility installations, the economy of relative fuel costs is usually the decisive factor. More recently, however, constraints associated with the need to limit the emission of air pollutants have reinforced this trend. The most recent concern, of course, is obtaining the necessary amount of any fuel at any price. This has fostered an intense interest in substitute fuels, that is, those produced by such processes as the gasification of coal and the sweetening of high sulfur crude oils.

The types, quantities and compositions of fuels consumed in stationary combustion processes reveal much about the emissions or at least the potential emissions from these sources. A number of organizations tabulate statistics on fuel usage.[47,48,49] The Federal Power Commission's annual summaries provide additional information on the quantities of fuel consumed in utility boilers and estimates of the emissions of particulate matter, sulfur dioxide and nitrogen oxides.[50] The emissions are calculated on the basis of emission factors for the various types of furnaces and the fuels consumed.

Natural Gas Combustion

Natural gas is a clean fuel, that is, it contains little sulfur or ash relative to those quantities present in oil and coal. In stationary sources, it is burned, almost without exception, in diffusion flames. The most challenging problem in burning gas is minimizing the formation of nitric oxide. Considerable attention has been devoted to the design of gas-fired furnaces to accomplish this.[51] The relative merits of horizontal and tangential firing as well as the location and spacing of the individual diffusion flames have been investigated. Off-stoichiometric or two-stage combustion arrangements have been used to limit the temperatures achieved in the primary combustion zone and at the same time maintain high combustion efficiencies.[51,52] Other variables have also been investigated. These include the effects of recirculating combustion products and the effect of the rate of mixing of secondary air with the primary combustion products on the formation of nitric oxide.[51,52]

In all fossil-fueled boilers, the combustion process is inherently non-adiabatic, that is, heat transfer from the flame to the surroundings is desirable. This is a favorable situation with respect to limiting NO formation since the greater the heat transfer from the flame, the lower its temperature and the less nitric oxide that will be formed, all other things being equal. Of course, this presupposes that radiant heat transfer occurs from the flame itself to the boiler surfaces rather than only from the hot combustion products to the boiler surface. If the furnace and the boiler surfaces are arranged so that they do not "see" one another, then the nonadiabatic argument does not apply and the combustion may be more adiabatic.

Oil Combustion

The term *oil* includes a wide variety of petroleum-based substances with differing physical properties and chemical compositions. For the most part, these substances are obtained by the refining of crude oil.

The composition of crude oil depends upon its geographic origin. An example of this variability is the organic sulfur content of crudes.[53] Refining processes can alter the distribution of sulfur, that is, the relative amounts present in the distillate and residual fractions obtained. Furthermore, sweetening or desulfurization processes can reduce the total sulfur present.[45, 54, 55, 56] In general, approximately 98% of whatever sulfur is present in a fuel is converted to SO_2, 1% to SO_3 and the remaining 1% resides in the ash.[57]

Organically bound nitrogen is present in some fuels and will contribute to NO formation.[52] Numerous other trace substances such as vanadium and other metals may also be present. In some cases they were originally present in the crude oil, while in other cases their presence was artificial insomuch as they were added to the fuel to decrease particulate emissions, to promote ignition of deposits on boiler surfaces, and to neutralize the ash.[44, 58]

The physical properties of a liquid fuel are important and affect both the nature of the fuel-burning equipment and the emissions. Light distillates are easier to burn than the heavier residuals that contain large amounts of carbonaceous residue, which are difficult to vaporize and burn. These are primarily used in larger combustion units.

Liquid fuels are generally atomized to facilitate combustion. There are a number of commonly used techniques[57, 60, 61] including high pressure steam or air, mechanical, centrifugal (rotary cup), and sonic atomization. The atomization process has an important bearing on the formation of trace species. Droplet size and spray pattern affect the rate of combustion and hence NO formation.[52] Furthermore, poor atomization can result in excess particulate formation.[62, 63] In many applications it is necessary to provide good atomization over a wide range of fuel flow rates. Sometimes this range is expressed in terms of the turndown ratio. The quality of atomization in a combustor or furnace (hence, particulate formation, etc.) may be excellent under some conditions but degenerate at higher or lower rates of fuel delivery.

Where lighter oils are burned, there is less chance of thermal decomposition, and vaporizing burners may be used.[57] In either case, *i.e.*, whether the fuel is atomized or vaporized, if the body of the fuel nozzle achieves a temperature sufficient to thermally decompose the fuel, internal deposits will form and upset the combustion process.

A number of variables influence the combustion process and emissions when burning liquid fuels. Two-stage combustion[57] has been used to limit flame temperature and hence NO formation. The nature of firing (tangential, horizontal, etc.)[57] and recirculation[64] are important design variables.

Vanadium oxide and iron-bearing ash can promote the oxidation of SO_2 to SO_3 in the effluent.[57] Since liquid fuels may contain significant quantities of sulfur, condensation to form an acid mist may occur at elevated temperatures. When this is a problem, it is desirable to maintain the effluent gases above their dew point to minimize corrosion of the effluent gas-handling equipment. The ash resulting from oil combustion may be quite hydroscopic and tend to cement to surfaces. This can be quite troublesome, particularly when the effluent is passed through certain types of stack gas treatment equipment.[65]

Emissions factors [particulate (size and chemical composition), CO, PNA's, acid smut] for oil combustion are summarized in References 57 and 66.

Future developments in the oil-fired furnaces may include high pressure (sometimes referred to as supercharged) units. These may operate at several atmospheres pressure and among other things have the advantage of much smaller volume. The volume of ancillary control equipment is reduced also. As progressively more stringent emission controls increase the complexity of combustion systems, this reduction in size of the boiler and its attendant control equipment may prove to be particularly attractive in some applications.

Coal Combustion

The chemical composition of coal, like that of oil, varies considerably with geographic origin. Coals are classified or ranked as anthracitic, bituminous, subbituminous and lignitic. Each of these ranks is further divided into different groups, and even then there is considerable room for variation within each subgroup. From an emissions point of view, perhaps the most important components are the sulfur, the volatile matter and the ash. Calorific value and moisture content are also important. The composition of coals (ultimate and proximate analysis) is discussed in Reference 53. The occurrence of minor and rare elements is discussed in References 59, 67-71. The sulfur in coal may be both organic and inorganically bound.[45] The inorganically bound sulfur is more easily removed (crushing and washing processes, etc.). Organic nitrogen may be present in coal and represents a potential problem for stringent NO control. Large quantities of ash-forming substances may be present in coal (up to approximately 25% by weight). Some of this contaminant is free mineral matter (rocks, soil, etc.) resulting from the mining process and may be removed by washing processes. Other ash is more difficult to remove. Another problem is that coals absorb moisture and tend to produce smoke when burned.

Coke is the solid residue that results when the volatiles (tars, aromatics, etc.) are removed from coal. Metallurgical cokes are produced from coals that contain relatively low quantities of metallic components.

Recent concerns about air pollution have increased interest in processes to upgrade coal by deashing[72] and by gasification[73] to produce synthetic fuels essentially devoid of the undesired sulfur and ash. The combustion of coal is discussed in References 53 and 74. Furnaces are usually classified by the feed method. Traditionally, this includes underfeed stokers, traveling and chain grates and spreader stokers. More recent furnace designs include suspension firing of pulverized coal in cyclone furnaces[53, 74, 75] and in fluidized beds.[53, 75, 76, 77]

Emission factors for coal combustion can be found in References 53, 76, 78 and 79. Test methods are given in Reference 53. Emissions can be controlled in one or more of three ways: (1) removal of problem substances from the fuel (i.e., deashing and gasification which were noted earlier); (2) modification of controllable variables associated with the combustion process itself; (3) removal of undesirable substances from the effluent. In the context of large stationary boilers, this latter process is usually referred to as stack-gas treatment. At present, there are more than a dozen different processes under development for the removal of sulfur dioxide. Basically these involve either adsorption (dry), absorption (wet) or catalysis.[80-85] The selection of a particular process is difficult at present since many are still in the pilot plant stage. Undoubtedly, the feasibility and economics of disposing of the recovered material is an important factor, and the optimum decision may well vary from one geographic area to another and with the quantity of recovered material in a particular application.

As this book goes to press, a number of stack-gas processes will just be coming on stream. In essence, each will be a large chemical plant treating tremendous volumes of effluent gas. Most convert the gaseous SO_2 to a solid dispersed in a liquid (slurry). The possibility exists that a portion of this product may be entrained in the stack gases and consequently be emitted as a particulate.

Radically new combustion processes are in the research stage.[86] An example is the magnetohydrodynamic seeding of coal combustion for the generation óf electric power.[87, 88, 89] Alternate combustion energy sources are also in varying stages of development. These include nuclear power fusion, fuel cells, utilization of solar energy,[90] harnessing of geothermal energy[86] and even of the wind.[91] Nuclear power has widespread application today and its use for the central generation of electrical power will increase. As the need for and cost of power increases, some of the other processes noted above may find limited application in specialized situations.

Incineration

Incineration, or the combustion of waste substances, is one of a number of alternatives for waste disposal. Other alternatives include landfills, ocean dumping, deep well disposal, composting, pyrolysis, reclamation and recycling. Historically, the decisive constraint in selecting a disposal technique has been economic rather than technological in nature. A number of aspects of solid waste management are discussed in References 92 and 93. In addition, a review of literature is also available.[94]

Incineration converts solid and liquid wastes into gaseous and finely divided particulate substances that can be dispersed and carried away by atmospheric currents. As the restrictions on water pollution and solid waste disposal increase, incineration may become more attractive. This is particularly true in large urban areas where land for solid waste disposal (dumps and landfills) is limited. A review of literature dealing with municipal incineration is available[95] as is a listing of definitions used in incineration technology.[96]

The process of incineration is unique in that more variability in the fuel is encountered than in perhaps any other type of combustion process. Wide variations exist in the chemical composition of waste substances as well as in their state of aggregation, moisture content (which even changes with the weather), and heat content. Proximate analysis (moisture, volatiles, etc.) and ultimate analysis (elemental) of refuse are tabulated in Reference 95. Methods of predicting solid waste composition are available,[96] and the calorific values of various common wastes have been tabulated.[95]

Combustion of solid wastes usually occurs in beds of some sort. A number of variables influence the combustion process and hence the emission of trace species. Feed arrangements are discussed in References 95 and 98. Agitation of the bed of unburned and burning material may be desirable to prevent searing of the material. When this occurs, a hard shell of decomposition products is formed, which precludes oxidation of the remaining material. The influence of air distribution is discussed in References 95 and 99. The primary air is responsible for the major amount of oxidation, while the secondary air results in oxidation above the bed to prevent or reduce the emission particulate and odoriferous substances. The kinetics of particulate burnout and gaseous reactions such as the water-gas shift in the secondary combustion zone are discussed in Reference 100. The size and shape of the combustion chamber are also important. Wall quenching and flame extinction may occur if the gases impinge on the wall before secondary oxidation is complete.[95] The temperature of combustion, both primary and secondary, is of importance for several reasons.[101] It affects the formation of NO_x, the fusing of ash, the slagging of the incineration walls and the emission of odors. Maintaining desired and reasonably constant temperature in an incinerator is aggravated by the variations encountered in the calorific content of the fuel. Blending of wastes and auxiliary firing if the calorific content of the fuel is insufficient can both be used to advantage.

A number of materials present special problems when they are incinerated. The thermoplastics and some metals, notably aluminum, may melt and run beneath the grate (resulting in a fire in the wrong place). Resins may polymerize coating surfaces and plug nozzles in incinerators designed to atomize liquid wastes. Magnesium can react thermally with iron surfaces to form globules on the refuse grate, etc. Still other substances such as polyolefins and nitrocellulose may detonate rather than burn.

Another problem arises when the material being incinerated has an elemental composition that results in the formation of undesirable combustion products.[102, 103] For example, polyvinyl chloride results in the formation of HCl and the freons (often used as a propellant in aerosol cans) can produce HF and in some cases HCl. Polyamides can form ammonia and amines. Styrene and asphaltine substances may result in the formation of dense smoke. Potential combustion products of plastics can be predicted from elemental analyses. Elemental analyses for the polymeric substances commonly encountered are given in Reference 104. Some substances contain sulfur[105] that will be converted to SO_2. Still other substances such as sludge contain phosphorus and heavy metals.[106] Paper may result in the emission of ash flakes. Tree leaves and grass contain organic nitrogen that can be converted to NO. They also contain other volatile organics (polysteroids, fatty acids, esters, lactones), some of which may be biologically active and may be emitted if these volatiles are not completely oxidized by combustion. The emission of bacteria associated with the incineration of fecal and food wastes is discussed in Reference 107.

Air pollution control equipment for municipal incineration is discussed in References 95 and 108. Operating and maintenance procedures are addressed in References 95 and 109. Test procedures for determining incinerator emissions are available.[95, 109, 110, 111] Finally, factors for the emission of various substances [including particles (weight, size distribution and chemical composition), hydrocarbons, aldehydes, PNA's, SO_2, NO, HCl, chlorine and phosgene] from municipal incinerators are available.[95, 102, 112]

Domestic and apartment house incinerators are smaller, simpler and less costly than municipal units. The design of apartment house incinerators is discussed in References 113 and 114. The "three T's"– temperature, time and turbulence–are important. Domestic and apartment house incinerators may be single or multiple chamber. They may be rather crude, such as the single flue incinerator in which fresh refuse drops on top of the burning pile and smothers combustion. Or, they may be more elaborate and contain automatic charging equipment and gas washers. The emissions from crude incinerators tend, of course, to be high and in some urban areas incineration of wastes in single and multiple dwelling units has been banned. Particulate emission factors for typical apartment house incinerators are tabulated in Reference 114. In designing domestic incinerators attention must be devoted to preventing the flue gas from entering the building either through the charging equipment or by downwash of the effluent and reentry through open windows or air conditioning intakes.

Refuse is sometimes incinerated by open burning. In the case of dumps, this may take place under intentional and controlled conditions or it may be accidental (spontaneous combustion, carelessness, etc.). Emission factors (carbon monoxide, hydrocarbons, aldehydes, organic acids, PNA's, NO, SO_2, NH_3) for open burning are tabulated in References 115 and 116. Bulky wastes (railroad cars, tires, appliances, furniture, trees and demolition lumber) that are difficult and costly to shred are often burned in the open.[117, 118] Open pit incineration techniques with forced air circulation and auxiliary fuel can help to insure complete combustion of these bulky wastes.[118] Agricultural wastes (grass, stubble, prunings, etc.) are often burned following the harvest. This is done for a variety of reasons, including controlling plant diseases, promoting the quick return of nutrients to the soil, and eliminating surface organic matter, which utilizes nitrogen during its decomposition. Emission factors for agricultural burning are contained in Reference 119.

The incineration of industrial process wastes is often accomplished in equipment that is specially designed for the particular waste stream.[102, 120] When the waste is a sludge or slurry, rotary kilns may be used so that an impervious skin will not form on the particles and retard their combustion. In other applications, such as incineration of lumbering and sawmill wastes, rather specific combustion devices such as the Tepee or Wigwam burner are used.[121] The incineration of liquid wastes, which may range from light volatiles to gummy organics with suspended waxy solids and mixtures thereof, is discussed in Reference 122. Gaseous wastes are produced by a number of common industrial processes. Solvent air mixtures result from drying, coating, spray painting, petroleum refining operations and venting of storage tanks. These may be oxidized thermally[95] in flares or in specially designed afterburners. The recovery of HCl from the incineration of chlorine-containing wastes by the use of a scrubbing system is discussed in Reference 123. Gaseous wastes can also be oxidized catalytically.[125, 126]

The increasing interest in incineration as a solid waste management technique has fostered a number of investigations for new and improved techniques.[126, 127] Among these is the use of fluidized beds[128] and waste heat recovery. Waste heat recovery has been practiced in Western Europe where fuel costs have been high. While there are few installations in the United States, interest is presently growing due to the increasing costs of fuel. Waste heat recovery may be accomplished by the installation of steam raising equipment in the combustion zone. The steam generated is used either to generate electric power or for space heating. Integrated applications such as this usually require a continuous combustion process. This may necessitate the installation of auxiliary firing equipment to generate steam when the incineration feed is interrupted. An alternate method of waste heat recovery is to utilize the combustion products generated by high pressure incineration to drive a turbine for the generation of electric power.

Still other novel methods for incineration are under development. One of these is underground burning. The material to be incinerated is covered with dirt fill and oxygen is pumped below the surface to facilitate combustion. Large volume reductions on the order of 90% can be achieved; at the same time, the residue is suitable as a landfill if left *in situ*.

For the most part, there is no recovery of resources from incineration. Two additional solid waste management techniques, wet oxidation and pyrolysis, are currently being developed that will permit significant resource recovery.

Wet oxidation involves the controlled oxidation of a substance to convert it to a useful and desired product.[108, 129, 130] The chemical reaction generally takes place in a slurry. The pressure and temperature are controlled to achieve the desired product distribution. This is an example of a closed system process as shown in Figure 76.

Pyrolysis of wastes also occurs in a closed system. Pyrolysis, however, involves the thermal decomposition of wastes in the absence of air.[131,132] Temperatures utilized are generally in the range of 1100°–1800°F but may exceed 2000°F. The products of pyrolysis include impure carbon (char), which can be used as a fuel for the pyrolysis itself or for other processes, light oils, pyroligneous (organic) acids, alcohols and gaseous hydrocarbons. Pyrolysis has a number of advantages. First, only the combustion products of the fuel needed to sustain the pyrolytic reactions are exhausted to the atmosphere. Second, marginally burnable materials can be consumed. Third, some substances such as metals can be separated and recovered. Finally, there is potential for the recovery of other chemicals in a usable form.

Industrial Processing

There are many different types of industrial processes which involve combustion. Combustion may be employed for the on-site generation of process steam and/or electric power, which in turn is used in a manufacturing process. On-site incineration of waste products may also be involved. These subjects have been addressed in the previous discussion of stationary boilers and of incineration. There are, however, still other combustion-related processes that have not been discussed, for example, metallurgical processes such as smelters and blast furnaces.

Due to the diverse kinds of industrial processes, no attempt will be made here to discuss the details of each. Rather, some review articles and literature surveys, which the reader should find of value in locating more specific information, will be identified. A compilation of emission factors has been published, including discussions of emissions from the chemical processing, metallurgical, mineral products, food and agricultural, petroleum and wood processing industries.[133] It should, of course, be recognized that not all of the emissions from industrial processes are combustion-related. Some may arise from simple entrainment of particles or the evaporation of liquids that come into contact with the atmosphere itself or with air used in the process. Each process is unique, and it is necessary to examine the various unit operations to conclude which emissions have combustion geneses and which do not.

Information on air pollutants resulting from the following processes and industries will be found in the references cited: iron and steel production,[134, 135] nonferrous metallurgical operations,[136-139] nonmetallic mineral products,[140] petroleum,[141, 142] inorganic chemicals,[143] pulp and paper,[144] and the food and feed industries.[145]

Finally, it should be noted that unique potential air pollution problems are created in industrial situations by virtue of the fact that large quantities of flammable materials are collected in one location. Industrial chemicals may contain high elemental concentrations of sulfur, phosphorus, nitrogen and halogens. Combustion of the same results in the formation of their respective oxides. Industrial processes, particularly in the chemical industry, often contain material under pressure. A failure of processing vessels or piping can release material either to the atmosphere or to an emergency holding system. Ignition of stored raw materials and products is also a potential problem as is their release and ignition during transport by truck, rail and vessel.

5.2 MOBILE SOURCES

Mobile sources of air pollutants include automobiles, trucks, buses, aircraft, vessels, snowmobiles and lawnmowers, to mention just a few. A more complete tabulation was given in Table I of Chapter 1. The great bulk of these sources are powered by either spark ignition, compression ignition or gas turbine internal combustion engines. The pace of developments in emission control technology, particularly in the case of the conventional spark ignition automobile engine, has been accelerated rapidly by recent legislation.[13]

Introductory texts that treat the subject of internal combustion engines are available.[146, 147] However, portions of the material in these texts are dated by recent developments in emission control technology. Other treatises are available that deal exclusively with emission control of mobile sources.[148, 149, 150] In addition the Society of Automotive Engineers has published three volumes containing selected papers on the subject.[151, 152, 153]

A number of analyses of the state of automotive emission control technology have been made.[154-159] These contain estimates of the emission control levels achievable by current and potential future power plants. Research and development is still in progress. Current information can be obtained from transcripts and proceedings of hearings on the subject that are conducted periodically by various legislative groups.[160]

The emissions from a mobile source are a strong function of the way in which the source is used or operated. They also depend upon other factors such as the environment (*i.e.*, ambient temperature, altitude, pressure, etc.) in which the source is operated. Consequently the procedures used to determine the emissions from mobile sources have an important bearing on the values obtained. In the case of the automobile, the test procedures and driving cycles have evolved over a period of time.[161, 162] This makes it difficult sometimes to compare the results of different studies.

The bulk of the measurements of vehicle emissions include only those species subject to legal constraints: HC, CO and NO_x. Data on the emission of other species such as the carbonyls, particulates and PNA's are sparse. Emission factors for mobile sources are reported in Reference 133. Data on the emissions from vehicles in use are also available.[163, 164, 165] Emissions of heavy duty (6000# and above) vehicles have been reported[166] as well as emissions from motorcycles,[167] other small engines,[168] locomotive Diesel engines and their marine counterparts[169] and aircraft.[170, 171]

The remainder of this section will be devoted to discussion of spark-ignited reciprocating engines, compression-ignited or Diesel reciprocating engines, gas turbines and miscellaneous alternate power plants. This latter category includes the more promising of the alternate engine approaches, the stratified charge engine, the Wankel engine and external combustion engines based on the Rankine and Stirling cycles.

Spark Ignited (SI) Reciprocating Internal Combustion Engine

During the early part of the century, automobiles were built with steam and electric as well as SI power plants. As a result of customer preferences, mostly related to cost, performance, range and start-up convenience, the SI engine emerged as the engine most universally used in automobiles. It is also used to power some trucks and buses, light aircraft and smaller vessels. Most of the larger engines of this type are of the four-cycle variety. Some of the smaller ones, which are used for motorcycles, outboard motors, chain saws, lawnmowers, minibikes, etc., are two-cycle engines.

Historically, the factors mentioned above, economics and durability (performance, warm-up characteristics, etc.) were the major constraints on the evolution of SI power plants. More recently, reduction of exhaust emissions and improvement of fuel economy[172] have become important.

Briefly, when public concern about air pollutants was first translated into legal constraints on the emission of hydrocarbons and carbon monoxide, the initial response of motor vehicle manufacturers was to modify the engine. Combustion chamber design was altered to improve combustion efficiency. Ignition timing and mixture stoichiometry were modulated to decrease emissions. In some cases, air was injected into the exhaust manifold to promote oxidation of partial combustion products. As the legal constraints became more severe, still other design modifications were necessary. NO_x control was achieved by increasing the valve overlap, reduction of the compression ratio, retarding the ignition timing, and recycling some of the exhaust products. The net result of all these changes has been to decrease exhaust emissions at the expense of increasing the fuel consumption and reducing the performance and driveability of the vehicle.

The addition of secondary combustion devices has been the focus of much investigation. The problems associated with the precise metering of the small amounts of fuel that are required for proper functioning of afterburners eliminated this approach. Thermal reactors, particularly when used in conjunction with rich engine combustion, can dramatically reduce the emissions of hydrocarbons and carbon monoxide. However, this reduction is achieved at the expense of even further decreases in fuel economy. Furthermore, due to the high temperatures generated, the reactor must be constructed of premium materials if it is to be durable. Oxidation catalysts show the greatest potential for use with IC engines. Their operating temperatures are lower than is required in a thermal reactor. Also, they reduce emissions while at the same time allowing engine variables to be better optimized to obtain recovery of some of the present fuel economy loss. On the debit side, they are easily poisoned. Thus, they require the use of unleaded gasoline. They may also increase the emission of other substances such as sulfates, sulfuric acid mist and particles containing trace metals. Catalysts that can reduce NO are still in an early stage of development.

The remainder of this section will be devoted to a more thorough discussion of the control techniques mentioned above with particular emphasis on their effects on the emission of trace species.

A carburetor is the most common device for metering the fuel. From the carburetor, the air-fuel mixture is conducted through the intake manifold to the cylinders. Maldistribution is said to occur when the quantities of air and fuel delivered to each cylinder vary in a way so that some cylinders are richer or leaner than others. Maldistribution limits the degree to which an engine can be leaned out. The overall mixture cannot be so lean that the lean-flammability limit is exceeded in any one cylinder or misfire will occur with an attendant increase in hydrocarbon emissions and loss in performance.

A choke is provided to assist in cold starting. It introduces a pressure drop upstream of the throttle plate in the carburetor. This reduces the pressure in the intake manifold and promotes vaporization of the more volatile fuel components. The effect of fuel volatility on emissions is discussed in References 173 and 174. During cold starting a pool of liquid fuel collects on the floor of the intake manifold. This aggravates the maldistribution problem. Heating of the intake air by exchange with the hot exhaust manifold can assist in vaporization of the fuel, as can a heat riser or "hot spot" on the floor of the intake manifold.[175]

The hydrocarbon and carbon monoxide emissions due to the abnormally rich combustion and the maldistribution problems that exist during the cold start period have taken on greater importance as the stringency of emission control has increased. Electrically modulated chokes have been used to provide more repeatable operation. Additives can also be used to keep the fuel metering system clean and thereby preserve the desired mixture ratio.

Fuel puddles in the intake manifold can also form during wide open throttle (rapid acceleration) operation due to the high absolute pressures in the intake manifold. To provide maximum power, carburetors provide mixture enrichment during such conditions. This, however, can increase carbon monoxide emissions. Another characteristic of carburetors that can cause mixture enrichment is operation at high altitudes (reduced ambient pressures). Altitude compensation can be provided to prevent this. Mixture enrichment may have little effect on a conventional SI engine other than increasing its emissions and perhaps decreasing its fuel economy and maximum power output. The consequences of mixture enrichment may be much greater on a vehicle equipped with catalysts or thermal reactors. Greater concentrations of combustibles enter the secondary combustion device. These, in turn, can produce destructively high temperatures as they oxidize. Consequently advanced control systems require fuel metering systems, which provide more precise fuel metering under all operating conditions.

Fuel injection is one alternative to carburation. Low pressure injectors, which introduce the fuel into the intake manifold just upstream of the intake valve, are sometimes used on SI engines. Fuel injection can provide more precise fuel metering. Electronic control systems can include sensors to measure manifold pressure and temperature, engine temperature, RPM, etc. and optimize the fuel flow for each particular operating condition. These electronic fuel injection (EFI) systems have been in production for a number of years on some European automobiles. In smaller vehicles, they can increase the power output, increase fuel economy for higher speed driving, and improve driveability particularly with manual shift transmissions.[176] Increasing concern about fuel economy and exhaust emissions coupled with a reduction in the cost of EFI and the trend toward electronic control of other engine functions will improve the outlook for EFI on larger vehicles. Also, sensors that measure oxygen concentration in the exhaust gases are under development. A feedback signal from these sensors has the potential of providing sufficiently rapid and precise control of the engine mixture to allow the use of a three-way catalyst.[177] These catalysts provide some degree of NO_x as well as HC and CO control. This, in turn, will make it possible to recover some of the losses that would accompany NO_x control by other methods.

In an SI engine, combustion occurs intermittently in each of the cylinders. The principles of combustion in SI engines are discussed in References 178 and 179. Thermal quenching of the layer of fuel-air mixture adjacent to the combustion chamber walls prevents or inhibits the reaction of the fuel contained in this layer.[180,181] Trapping of an additional portion of the mixture in crevices alongside the piston also prevents the flame from oxidizing this portion of the mixture. The hydrocarbons in the quench layer adjacent to the wall and those present in crevices are not completely unburned. However as a result of the temperatures to which they are exposed, they do undergo precombustion reactions. Hydrocarbon emissions are also affected by the surface-to-volume ratio of the combustion chamber. Initial attempts to control hydrocarbon emissions were in part based on modifying the shape of the combustion chamber to reduce the fraction of the charge that survived. There are still other phenomena that also influence hydrocarbon emissions. Deposits on the surfaces of the combustion chamber may absorb hydrocarbons during compression, store them during the combustion event and then release them during the expansion process. Burning of oil due to poor mechanical condition of the engine increases hydrocarbon emissions. An approach that has been used to reduce the hydrocarbons exhausted is retarding the ignition. Late ignition of the charge results in burning during the expansion stroke. This in turn results in high temperatures of the exhaust gas exiting from the combustion chamber. If the mixture is lean, the oxygen in the exhaust gases will react with hydrocarbons in the exhaust manifold. If the mixture is rich, air must be injected just downstream of the exhaust valve to promote this oxidation.

Thermodynamic and kinetic models are available for analyzing the high temperature combustion which, with the exception of the quench zone, occurs throughout the combustion chamber.[183-188] Carbon monoxide and nitrogen oxides are formed in this region. The concentration of carbon monoxide is, of course, primarily a function of the mixture stoichiometry. Nitrogen oxide formation depends upon the peak temperatures that are achieved during combustion. Peak flame temperatures can be reduced by utilizing a rich or lean mixture, by internal or external recirculation of exhaust gases, and by retarding the ignition.[189,190] Other techniques, such as water injection, are theoretically sound but appear impractical.[191,192]

As noted earlier combustion in a SI engine is intermittent. Variations occur from cycle to cycle in the same cylinder and from cylinder to cylinder.[193,194,195,196] These variations, which are sometimes referred to as cyclic dispersion, are related to a number of factors including the maldistribution discussed earlier, variations in ignition timing, imperfect mixing of the fresh and residual gases, and variations in combustion chamber turbulence. The importance of cyclic dispersion on exhaust emissions should not be underestimated. Not only can the hydrocarbons rise as incipient misfire occurs in one or more cylinders, but an inordinate amount of NO may be produced by those cylinders in which higher peak temperatures are achieved.

The importance of achieving optimum combustion through more precise control of ignition timing (placement of the combustion event), amount of exhaust gas recirculated, etc., increases as emission control requirements become more stringent. Electronic modulation of these variables based on inputs from multiple sensors measuring both engine and ambient conditions becomes more attractive as the number of things to be optimized increases (e.g., exhaust emissions, feedstream composition to secondary combustion devices, fuel consumption, durability, etc.). Conventional mechanical systems that provide control based on vacuum and centrifugal sensors provide a degree of control that is at best a compromise. It is anticipated that the use of electronic systems that provide greater precision over a wider range of operating conditions will increase in the future.

The simplest type of secondary combustion process is created when the ignition is retarded to promote oxidation of hydrocarbons and carbon monoxide in the exhaust ports and manifold. Additional air may or may not be needed depending upon the stoichiometry of the fuel-air mixture entering the engine. The kinetics associated with thermal reactions in the exhaust have been studied.[197, 198, 199]

Normally, a thermal reactor is thought of as a device promoting the oxidation of unburned hydrocarbons and carbon monoxide by containing the exhaust gases at a high temperature for a sufficient period of time to allow the desired degree of reaction to occur. When secondary air is injected, the reactor must also promote the mixing of this air with the effluent gases from the combustion chambers. A number of reactors have been developed.[200, 201, 202, 203] Investigations of some of these systems have revealed the multiple operating point phenomena[199, 204] discussed in Section 3.2. Reactors can provide very effective control of hydrocarbons and carbon monoxide. They do not provide any reduction in NO_x. Simultaneous control of all three pollutants requires such rich combustion that extremely high temperatures are produced in the reactor when the resultant partial combustion products oxidize. Not only does this present serious reactor durability problems, but fuel consumption is increased substantially.

Oxidation catalysts have been used for some time to reduce the concentration of carbon monoxide in the exhausts of engines that are used in confined spaces (fork lift trucks in warehouses, power sources in underground mines, etc.). Currently, they are being developed for use on automobiles beginning with the 1975 and 1976 model years.[205, 206, 207] Oxidation catalysts are attractive because they not only produce dramatic reductions in total hydrocarbons in the exhaust but they also reduce the concentrations of olefinic and aromatic species that are photochemically reactive in the atmosphere. They reduce the latter proportionately more than they reduce the concentration of relatively unreactive methane and other light paraffins.[208, 209] Oxidation catalysts have also been shown to be effective in reducing the emissions of PNA's.[210, 211] These catalysts are subject to poisoning by the combustion products of tetra ethyl lead, which is a commonly used gasoline octane improver. Other additives such as phosphorus are also detrimental. If attrition of the catalyst occurs, the fine particles produced may be emitted as particulate matter. Also, there is evidence that oxidation catalysts can promote the conversion of sulfur dioxide into sulfates and sulfuric acid mist.

Catalysts that promote the reduction of nitric oxide in automotive exhaust gases are still in the early stage of development. Dual bed catalyst systems similar to those discussed in Section 3.4 have been studied.[212, 213] The production of ammonia by the reduction catalyst and its subsequent oxidation to nitric oxide by the oxidation catalyst reduces the net efficiency of the system. Ruthenium has been found to produce much less ammonia than other materials that also catalyze the NO reduction. However, when such catalysts are exposed to transient oxidizing conditions, the ruthenium is lost as a volatile oxide. Other durability problems have also been observed with reduction catalysts. As mentioned earlier, the use of an oxygen sensor in the effluent gases to provide precise control of the engine mixture may allow the use of "3-way" catalysts, which do provide a measure of NO_x reduction.

The use of cyclone separators has been proposed to provide particulate control.[200] Particles arise, in part, from additives in the fuel and the oil.[214, 215, 216] Regulation of fuel additives is another approach for reducing the emissions of particulate matter.

The influence of alternate fuels on exhaust emissions has been studied.[217, 218, 219] LPG has some unique fuel handling problems associated with its use, but it has been used as a fuel by some fleets and agricultural users. Since it is a gaseous fuel, maldistribution and manifold wetting problems are reduced or alleviated and low exhaust emissions can be achieved, particularly when the engine is cold. LNG has also been used as a fuel.[217] The fuel handling problems associated with cryogenic storage are more severe than with LPG.

The use of alcohol and alcohol-hydrocarbon mixtures have also been investigated.[220, 221, 222] Data are available on the emissions from pure hydrocarbons[223] and ammonia as a fuel.[224]

Finally, some additional sources of information on exhaust emissions on specific substances are as follows: particles,[225, 226] nitrogenous compounds,[227, 228] PNA's,[229] and oxygenates.[230, 231, 232]

Compression Ignition (CI) Reciprocating Internal Combustion Engines

Compression ignition (or Diesel) engines are used to power heavy trucks, buses, locomotives and vessels. They have found limited application in automobiles.[233] The reasons why Diesel and SI engines have essentially different markets are related to the different characteristics of these engines. Diesels have higher compression ratios and the pressures achieved in the combustion chambers are considerably greater than those achieved in an SI engine. Consequently, the Diesel engine is of much heavier construction and has a higher initial cost than the SI engine. The high pressure fuel injection system also contributes to the higher cost. The fuel economy achieved by Diesels is, however, superior to that achieved by the SI engine. Thus, Diesels are attractive in trucks and other commercial vehicles where high initial cost is less of a detriment and is offset by savings in fuel costs.

Diesel engines come in many configurations: 2- and 4-cycle engines, open and closed (prechamber) combustion chamber types, and with specially shaped pistons and inlet valve configurations to provide chamber turbulence (swirl) etc. The many different configurations will not be discussed in detail here other than to recognize that configuration does not affect the combustion process and exhaust emissions.[234, 235, 236, 237] Additional discussion of the emissions from Diesel engines is contained in References 149, 238, 239. Modeling of CI engines is discussed in Reference 240.

Smoke and odor are the most noticeable emissions of Diesel engines. White smoke is emitted when the engine is cold and consists largely of unburned or partially burned fuel droplets. It may be accompanied by lachrymatory substances such as carbonyl compounds and other partial oxidation products. Black smoke, by way of contrast, is emitted when the engine is operating at high load. This smoke consists of carbonaceous particles produced by the pyrolytic reactions that occur in the core of the fuel spray when large amounts of fuel are injected into the combustion chamber. The physical and chemical characteristics of particles in Diesel engine exhaust are discussed in Reference 243. PNA emissions are discussed in Reference 244. Substances can be added to the fuel to reduce smoke.[243] Turbochargers can reduce smoke at high loads by introducing additional air into the combustion chamber and thereby preventing excessively rich mixtures. The characteristic odor of Diesels may be associated with the fuel or with minute concentrations of species generated in the combustion process.[244-249] The excess air that is usually present in Diesel engines results in relatively low carbon monoxide and hydrocarbon concentrations. In the case of hydrocarbons, some of the species that are contained in Diesel exhaust have relatively high molecular weights, and special precautions must be taken to insure that losses do not occur in sampling.[250] The nitrogen oxides formed are strongly affected by combustion chamber design.[230]

Gas Turbines

Gas turbines are widely used to power aircraft, and they have also been considered for automotive, truck and bus applications. Very briefly, turbines operate by compressing the inlet air. A portion of this air then flows into the primary combustion zone of the combustor. The fuel is also introduced into this zone as either a spray of liquid droplets or as a gas. Combustion of the fuel and the primary air raises the temperature of the flowing gases. The remainder of the inlet air is introduced along the walls of the combustor. This secondary air has two functions. First, it provides additional oxygen to promote oxidation of any partial combustion products from the primary zone. Second, its flow is arranged so as to prevent direct impingement of the hot combustion products on the combustor wall. If such impingement did occur, not only would the cooler walls tend to quench the oxidation reaction but hot spots could develop on the walls leading to material failure. Following combustion, the hot gases expand as they flow through power stage of the turbine. The work produced by this expansion that is not used to drive the inlet gas compressor or to power engine accessories provides motive power for the craft or vehicle.

A large body of literature on gas turbines is available. The focus of this discussion will be on the emissions thereof. Emissions of large aircraft type turbines are reported in References 251, 252, 253 and 254. In addition, emissions measurements have also been made on smaller turbines of the type that might find application in ground based vehicles.[255-260] Additional information can be found on the effect of combustor design[261-263] and on fuel composition.[264]

Due to the excess of air in a turbine and its arrangement to prevent quenching of the hot combustion gases by the cooler walls of the combustor, the hydrocarbon and carbon monoxide emissions are normally quite low. Another aspect of turbine combustion that contributes to low emissions is that the combustion is more nearly steady state than is combustion in a reciprocating (SI or CI) engine. Smoke can be formed if very rich local air fuel ratios are achieved in the primary combustion zone. This is more likely to occur in aircraft turbines that operate over a large inlet pressure range. No smoke may be formed at high altitudes. However, during lower altitude operation (landing and takeoff), the higher densities of air in the combustor may upset the fuel spray pattern and result in a rich core that produces carbonaceous particles. Use of additional air to atomize the fuel (refer to Section 4.3) may reduce the richness of the core and hence the smoke formation. This may be accomplished at the expense of additional nitric oxide formation due to the greater availability of oxygen in the spray core.

Nitric oxide emissions from a turbine are largely dependent on the conditions prevailing in the primary zone of the combustor. It is necessary to control nitric oxide formation by combustor design. Local air fuel ratios and internal recirculation patterns should be regulated to prevent local hot spots within the combustor that could result in the formation of excessively large amounts of nitric oxide. Since the overall air-fuel ratio in a gas turbine combustor is lean, once nitric oxide is formed, significant decomposition cannot be expected. Catalytic reduction is not a possibility either. Not only is the concentration of reactants low but the effluent gases are strongly net oxidizing rather than net reducing. This latter statement also applies to other power plants that operate with an overall excess of air, the Diesel discussed previously as well as the stratified charge engines that will be discussed in the next section. The net oxidizing conditions that prevail in the effluents of these engines make the application of a dual-catalyst system, as outlined on the bottom of Figure 30, impractical. Too much secondary fuel would have to be added to consume the excess oxygen and produce net reducing conditions.

Miscellaneous

The Wankel or rotary SI engine has been used to power automobiles.[265, 266] The exhaust port configuration and the high concentration of combustibles in the effluent favor the use of a thermal reactor to control the emissions from this type of engine. The rotary engine offers a number of potential advantages. It has fewer parts and is lighter and smaller than its reciprocating counterpart. Its operation is also very smooth. On the debit side, thus far Wankels have exhibited poorer fuel economy than reciprocating engines and seal durability problems have also been experienced. Currently, much development work is in progress to evaluate the potential of this engine.[267, 268]

Another class of engines that is currently the subject of intense research efforts is the stratified charge engine. Stratification of the fuel-air mixture may be achieved by injecting the fuel directly into the combustion chamber[269, 270] or by the use of a divided or a prechamber.[271, 272] In either case, the potential of this engine for achieving low emissions is based on the idea of initiating the burning of the fuel in a rich zone where nitric oxide formation will be low, subsequently completing the combustion in a lean region where any remaining unburned fuel, partially oxidized hydrocarbons and carbon monoxide will be oxidized. In addition, there is potential for achieving excellent fuel economy and for burning a range of fuels.[273] A number of theoretical analyses of stratified combustion are available.[274, 275]

All of the engines discussed thus far are internal combustion engines. A number of external combustion engines have also been investigated.[276] Two external combustion engines that have received considerable attention are those based on the Rankine[277-282] and the Stirling cycles.[283-285]

Still other types of power plants have been proposed. Among them are the electric[154] and various hybrid power plants. The electric vehicle does eliminate the emission of combustion-related air pollutants from the source itself. However, the problem is transposed to the central power station where electricity is generated to charge the batteries of the electric vehicle.[286]

Judgment must be used in evaluating the claims made for some of the less well-known potential alternate power plants. In some cases, the claims are based only on "paper studies" and the emissions and efficiencies are no better than the assumptions made. In other cases, they may be based on limited experimental data from prototype vehicles, engine dynamometer studies or even bench scale testing of the combustor alone. The results of these tests may not be representative of the emissions or the efficiencies that might be achieved from production vehicles.

REFERENCES

1. Air Pollution Control Association, Pittsburgh, Pa., monthly journal, *J. Air Poll. Control Assn.*
2. The Combustion Institute, Pittsburgh, Pa., bimonthly journal, *Combustion and Flame.* Also organizes the biannual *Symposium (International) on Combustion* and publishes the proceedings.
3. American Petroleum Institute, Washington, D.C., publishes *Proc., Div. of Refining.*
4. American Society of Mechanical Engineers, publishes *Trans.*, and *J. Engrg. for Power.*
5. Institution of Mechanical Engineers, London, England, publishes *Proc.*
6. American Chemical Society, Washington, D.C., publishes *Environ. Sci. Technol.*
7. American Institute of Chemical Engineers, New York, N.Y., publishes *Jour.*
8. Society of Automotive Engineers, New York, N.Y., publishes *Jour.* and *Trans.*
9. *Comb. Sci. Technol.*, published monthly by Gordon and Breach Science Publishers, New York, London, Paris.
10. *Fuel—The Science of Fuel and Energy* published monthly by IPC Science and Technology Press, Ltd., England.
11. *J. Inst. of Fuel*, London.
12. *Staub Reinhaltung der Luft*, translated and published for the U.S. Environmental Protection Agency, Available from NTIS, U.S. Dept. of Commerce, Springfield, Va. 22151.
13. Clean Air Act (42 U.S.C. 1857 *et seq.*) includes the Clean Air Act of 1963 (P.L. 88-206), and amendments made by the "Motor Vehicle Air Pollution Control Act" — P.L. 89-272 (October 20, 1965), the "Clean Air Act Amendments of 1966" — P.L. 89-675 (October 15, 1966), the "Air Quality Act of 1967" — P.L. 90-148 (November 21, 1967), and the "Clean Air Amendments of 1970" — P.L. 91-604 — (December 31, 1970).
14. AP group provides information of general interest in the field of air pollution control. Copies can be purchased from the Government Printing Office (GPO), Washington, D.C. Copies of out-of-print reports no longer available from GPO can be purchased from National Technical Information Service (NTIS), Springfield, Va. 22151.
15. The APTD series is issued to report technical data of interest to a limited readership. Available from NTIS, Springfield, Va. 22151.
16. Air Pollution Technical Information Center, Office of Air and Water Programs, Environmental Protection Agency, Research Triangle Park, N.C. 27711. Also prepares Air Pollution Abstracts which are issued monthly.
17. Bureau of Mines, U.S. Department of Interior, Washington, D.C., Conducts research, with particular emphasis on fuels, related to the combustion and emissions of both mobile and stationary sources. Publishes *Reports of Investigations.*
18. National Bureau of Standards, U.S. Department of Commerce, Washington, D.C. 20234. Active in the development of standard reference materials for

measurement of trace species emitted by sources and of their concentrations in the atmosphere. Also the Chemical Kinetics Information Center can provide kinetic data on elementary combustion reactions.

19. United Nations, New York, N.Y. Conducts various studies on the emissions from combustion related processes. Additional studies available from Economic Commission for Europe, Palais Des Nations, 1211 Geneva, Switzerland.

20. American Society for Testing Materials, 1916 Race Street, Philadelphia, Pennsylvania 19103. Publishes standard test procedures for analysis of fuels and measurement of source emissions.

21. "Air Pollution Manual—Part II, Control Equipment," American Industrial Hygiene Association. 25711 Southfield, Michigan 48075 (1968).

22. "Air Pollution Control," *Chem. Eng. (N.Y.), Deskbook Issue,* June 21, 1971, pp 131-141.

23. "Foundry Air Pollution Control Manual," 2nd Ed. American Foundrymens Society, Des Plaines, Ill., 1967.

24. Loffler, L. "Collection of Particles by Fiber Filters," in *Air Pollution Control, Part I,* W. Strauss, Ed. (New York: Wiley Interscience, 1971) chapter 6.

25. Iinoya, K. and C. Orr. "Source Control by Filtration," in *Air Pollution, Vol. III, Sources of Air Pollution and Their Control,* A. C. Stern, Ed. (New York: Academic Press, 1968) chapter 44.

26. Teller, A. J. "Air Pollution Control," *Chem. Eng. (N.Y.) Environmental Deskbook Issue* (May 8, 1972) pp 93-98.

27. Caplan, K. J. "Source Control by Centrifugal Force and Gravity," in *Air Pollution, Vol. III, Sources of Air Pollution and Their Control,* A. C. Stern, Ed. (New York: Academic Press, 1968) chapter 43.

28. White, H. J. *Industrial Electrostatic Precipitation* (Reading, Mass.: Addison Wesley, 1963).

29. Gottschlich, C. F. "Source Control by Electrostatic Precipitation," in *Air Pollution, Part III, Sources of Air Pollution and Their Control,* A. C. Stern, Ed. (New York: Academic Press, 1968) chapter 45.

30. Robinson, M. "Electrostatic Precipitation," in *Air Pollution Control, Part I,* W. Strauss, Ed. (New York: Wiley Interscience, 1971) chapter 5.

31. Calvert, S., D. Lundgren, and D. S. Mehta. "Venturi Scrubber Performance," *J. Air Poll. Control Assn.,* 22, 529 (1972).

32. Calvert, S. "Source Control By Liquid Scrubbing," in *Sources of Air Pollution and Their Control,* A. C. Stern, Ed. (New York: Academic Press, 1968) chapter 46.

33. First, M. W., and F. J. Viles. "Cleaning of Stack Gases Containing High Concentrations of Nitrogen Oxides," *J. Air Poll. Control Assn.,* 21, 122 (1971).

34. "Basic Technology—Air Pollution Control," *Chem. Eng. (N.Y.), Environmental Deskbook Issue* (April 27, 1970) pp. 165-172.

35. Turk, A. "Source Control by Gas-Solid Adsorption and Related Processes," in *Air Pollution, Vol. III—Sources of Air Pollution and Their Control,* A. C. Stern, Ed. (New York: Academic Press, 1968) chapter 47.

36. Hanf, E. B. "Entrainment Separation Design," *Chem. Eng. Progr.,* 67, 11, 54 (1967).

37. Veldhuizen, H., and J. Ledbetter. "Cooling Tower Fog: Control and Abatement," *J. Air Poll. Control Assn.,* 21, 20 (1971).

38. Crocker, B. B. "Water Vapor in Effluent Gases: What to do about Opacity Problems," *Chem. Eng.* (July 15, 1968) p 109.

39. Engdahl, R. B. "Stationary Combustion Sources," in *Air Pollution, Vol. III, Sources of Air Pollution and Their Control,* A. C. Stern, Ed. (New York: Academic Press, 1968) chapter 32.

40. Stern, A. C. "Efficiency, Application and Selection of Collectors," in *Air Pollution, Vol. III, Sources of Air Pollution and Their Control,* A. C. Stern, Ed. (New York: Academic Press, 1968) chapter 42.

41. "Air Pollution Engineering Manual," J. A. Danielson, Ed. Environmental Protection Agency, AP-40 (May, 1973), 2nd Edition. See Reference 14 for additional information.

42. *Steam—Its Generation and Use* (New York: The Babcock and Wilcox Company, 1955).

43. Francis, W. *Boiler House and Power Station Chemistry,* 4th edition (London Edward Arnold, Ltd., 1962).

44. Engdahl, R. B. "Stationary Combustion Sources," in *Air Pollution, Vol. III— Sources of Air Pollution and Their Control,* A. C. Stern, Ed. (New York: Academic Press, 1968).

45. Strauss, W. "The Control of Sulfur Emissions from Combustion Processes," in *Air Pollution Control—Part I,* W. Strauss, Ed. (New York: Wiley-Interscience, 1971).

46. "Fossil-Fuel Boilers Face a Bright Future," *Energy Int.* (February, 1973) p 13.

47. "Minerals Yearbook," prepared by the staff of The Bureau of Mines and the U.S. Department of Interior. Available from the U.S. Government Printing Office, Washington, D.C.

48. Monthly surveys entitled "Fuel Oils by Sulfur Content" and "Petroleum Statement" are available from the U.S. Department of the Interior, Bureau of Mines, Washington, D.C. 20240.

49. "A Staff Report on Monthly Reports of Cost and Quality of Fuel for Steam-Electric Plants," Bureau of Power, Federal Power Commission. The first monthly report is for the month of April, 1973. Based on FPC Form 423 and tabulates only statistics for plants of 25 MW and greater capacity. Monthly news releases are also available for all plants and are based on Form FPC-4.

50. "Steam Electric Plant Air and Water Quality Control Data—for the Year Ended December 31, 1969—Summary Report," Bureau of Power, Federal Power Commission, Washington, D.C. 20426. This is the first of a series of yearly summary reports that FPC will issue. It is based on FPC Form 67 and covers only plants of 25 MW or greater capacity.

51. Hall, H. J., and W. Bartok. "No$_x$ Control from Stationary Sources," *Environ. Sci. Technol.,* **5,** 320 (1971).

52. Bagwell, F. A., K. E. Rosenthal, D. P. Teixeira, B. P. Breen, N. Bayard de Volo, and S. Kerho. "Utility Boiler Operating Modes for Reduced Nitric Oxide Emissions," *J. Air Poll. Control Assn.,* **21,** 702 (1971).

53. Smith, W. S., and C. W. Gruber. "Atmospheric Emissions from Coal Combustion—An Inventory Guide," U.S. Department of Health, Education and Welfare, Public Health Service Publication, 999-AP-24, April, 1966. See Reference 14 for additional information.

54. Cortelyou, C. G., R. C. Mallatt, and H. H. Meredith. "A New Look at Desulfurization," *Chem. Eng. Prog.,* **64,** 53 (1968).

55. Spaite, P. W., and R. P. Hangebrauck. "Pollution from Combustion of Fossil Fuels, Air Pollution—Part I," U.S. Government Printing Office, 1970.

56. Finfer, E. Z. "Some Technical and Economic Aspects of Residual Fuel Oil Desulfurization," *J. Air Poll. Control Assn.,* **15,** 485 (1965).

57. Smith, W. S. "Atmospheric Emissions from Fuel Oil Combustion—An Inventory Guide," U.S. Department of Health, Education and Welfare Publication No. 999-AP-2 (1962). See Reference 14 for additional information.

58. Finfer, E. Z. "Fuel Oil Additives for Controlling Air Contaminant Emissions," *J. Air Poll. Control Assn.,* **17,** 43 (1967).

59. Lee, R. E., and D. J. von Lehmden. "Trace Metal Pollution in the Environment," *J. Air Poll. Control Assn.,* **23,** 853 (1973).

60. Peskin, R. L., and R. J. Raco. "Analytical Studies on Mechanisms of Fuel Oil Atomization," American Petroleum Institute Publication No. 1727 (1967).

61. Perron, R. R., J. E. McCullough, and E. S. Shanley. "The Development of Ultrasonic Atomizer for Domestic Oil Burners," American Petroleum Institute Publication No. 1725-A, 1967.

·62. McGarry, F. J., and C. J. Gregory. "A Comparison of the Size Distribution of Particles Emitted from Air, Mechanical, and Steam Atomized Oil-Fired Burners," *J. Air Poll. Control Assn.*, **22**, 636 (1972).

63. Molino, N., and E. P. Zabolotny. "Operational Method for Reduction of Particulate Emissions from an Oil Fired Utility Plant," presented at the Central States Section of the Combustion Institute, Champaign, Ill. (March 1973).

64. Cooper, P. W., R. Kamo, C. J. Marek, and C. W. Solbrig. "Recirculation and Fuel-Air Mixing as Related to Oil Burner Design," American Petroleum Institute Publication No. 1723, 1964.

65. Bagwell, F. A., and R. G. Velte. "New Developments in Dust Collecting Equipment for Electric Utilities," *J. Air Poll. Control Assn.*, **21**, 781 (1971).

66. Wasser, J. H., R. P. Hangebrauck, and A. J. Schwartz. "Effects of Air-Fuel Stoichiometry on Air Pollutant Emissions from an Oil-Fired Test Furnace," *J. Air Poll. Control Assn.*, **18**, 332 (1968).

67. Schlesinger, M. D., and H. Schultz. "An Evaluation of Methods for Detecting Mercury in Coals," U.S. Bureau of Mines Report of Investigation No. 7609, 1972.

68. Abernethy, R. F., and F. H. Gibson. "Rare Elements in Coal," U.S. Dept. of Interior, Bureau of Mines Information Circular 8163, October 1962.

69. Chow, T. J., and J. L. Earl. "Lead Isotopes in North American Coals," *Science*, **176**, 510 (1972).

70. Bayliss, R. J., and A. M. Whaite. "A Study of the Radium Alpha-Activity of Coal Ash and Particulate Emission at a Sydney Power Plant," *Air Water Poll.*, **10**, 813 (1966).

71. Billings, C. E., A. M. Sacco, W. R. Matson, R. M. Griffin, W. R. Coniglio, and R. A. Harley. "Mercury Balance on a Large Pulverized Coal-Fired Furnace," *J. Air Poll. Control Assn.*, **23**, 773 (1973).

72. Brant, V. L., and B. K. Schmid. "Pilot Plant for De-Ashed Coal Production," *Chem. Eng. Prog.*, **65**, 55 (1969).

73. Chopey, N. P. "Coal Gasification: Can It Stage a Comeback?," *Chem. Eng.* (April 3, 1972) p 44.

74. Cuffe, S. T., and R. W. Gerstle. "Emissions from Coal-Fired Power Plants—A Comprehensive Summary," U.S. Dept. of Health, Education and Welfare Publication No. 999-AP-35, 1967.

75. Ehrlich, S. "Air Pollution Control Through New Combustion Processes," *Environ. Sci. Technol.*, **4**, 396 (1970).

76. Frankel, R. J. "Problems of Meeting Multiple Air Quality Objectives for Coal-Fired Utility Boilers," *J. Air Poll. Control Assn.*, **19**, 18 (1969).

77. Browning, J. E. "Fluidized Combustion Broadens Its Horizons," *Chem. Eng.* (June 28, 1971) p 44.

78. Gerstle, R. W., S. T. Cuffe, A. A. Orning, and C. H. Schwartz. "Air Pollutant Emissions from Coal-Fired Power Plants, Report #2," *J. Air Poll. Control Assn.*, **15**, 59 (1965).

79. Lee, R. E., and C. F. Smith. "Size Distribution of Suspended Particles from Lignite Combustion," *Environ. Sci. Technol.*, **6**, 929 (1972).

80. "SO₂ Removal Technology Enters Growth Phase," *Environ. Sci. Technol.*, **6**, 688 (1972).

81. Slack, A. V. "Air Pollution: The Control of SO₂ from Power Stacks, Part III—Process for Recovering SO₂," *Chem. Eng.* (December 4, 1967) p 188.

82. Squires, A. M. "Air Pollution: The Control of SO₂ from Power Stacks," *Chem. Eng.* (November 20, 1967) p 133.

83. "Outside U.S., Tough Laws Spur SO₂ Removal Efforts," *Chem. Eng.* (November 4, 1968) p 84.

84. Bienstock, D., J. H. Field, S. Katell, and K. D. Plants. "Evaluation of Dry Processes for Removing Sulfur Dioxide from Power Plant Flue Gases," *J. Air Poll. Control Assn.*, **15**, 459 (1965).

85. Kiyoura, R., and M. Munidasa. "Available Desulphurization Technology and Its Application to the Rational Utilization of Fossil Fuel," *Proc. Second International Clean Air Congress* (New York: Academic Press, 1971) pp 842-850.
86. Weaver, K. F. "The Search for Tomorrow's Power," *National Geographic,* **142,** 650 (1972).
87. Bienstock, D., R. J. Demski, and R. C. Kurtzrock. "High Temperature Combustion of Coal Seeded with Potassium Carbonate in the MHD Generation of Electric Power," U.S. Bureau of Mines Report of Investigation 7361, 1970.
88. Bienstock, D., P. D. Bergman, J. M. Henry, R. J. Demski, J. J. Demeter, and K. D. Plants. "Air Pollution Aspects of MHD Power Generation," presented at the 13th Symposium, Engineering Aspects of Magnetohydrodynamics, Stanford Univ., March 1973.
89. Bergman, P. D., and D. Bienstock. "Mixed Potassium-Cesium Seeding in Open-Cycle MHD Power Generation," presented at 13th Symposium Engineering Aspects of Magnetohydrodynamics, Stanford Univ., March 1973.
90. An Assessment of Solar Energy as a National Energy Resource," prepared by the NSF/NASA Solar Energy Panel, December, 1972.
91. McCall, J. "Windmills," *Environment,* **16,** 6 (1973).
92. Baum, B., and C. H. Parker. *Solid Waste Disposal, Vol. I, Incineration and Landfill* (Ann Arbor, Michigan: Ann Arbor Science Publishers, 1973).
93. Baum, B., and C. H. Parker. *Solid Waste Disposal, Vol. II, Reuse/Recycle and Pyrolysis* (Ann Arbor, Michigan: Ann Arbor Science Publishers, 1974).
94. "Solid Waste Management: A List of Available Literature," U.S. Environmental Protection Agency, October, 1972. Solid Waste Management Series Publication SW-58.16.
95. Stear, J. R. "Municipal Incineration—A Review of Literature," U.S. Environmental Protection Agency, AP-79, June, 1971.
96. Corey, R. C. "Definitions of Terms used in Incinerator Technology," *J. Air Poll. Control Assn.,* **15,** 125 (1965).
97. Boyd, G. B., and M. B. Hawkins. "Methods of Predicting Solid Waste Characteristics," U.S. Environmental Protection Agency, Solid Waste Series, SW-23c (1971).
98. Stenburg, R. L., R. P. Hangebrauck, D. J. von Lehmden, and A. H. Rose. "Effects of High Volatile Fuel on Incinerator Effluents," *J. Air Poll. Control Assn.,* **11,** 376 (1961).
99. Stenburg, R. L., R. P. Hangebrauck, D. J. von Lehmden, and A. H. Rose. "Field Evaluation of Combustion Air Effects on Atmospheric Emissions from Municipal Incinerators," *J. Air Poll. Control Assn.,* **12,** 83 (1962).
100. Huntington, R. L. "Incinerator Afterburner Particulate Emission Factors: A Correlation and Combustion Study," presented at the 66th Annual Meeting of the Air Poll. Control Assn., Chicago, Ill. (June 1973).
101. Meissner, H. G. "Incinerator Temperature—How to Calculate and Control It," *J. Air Poll. Control Assn.,* **11,** 479 (1961).
102. Carotti, A. A., and E. R. Kaiser. "Concentrations of Twenty Gaseous Chemical Species in the Flue Gas of a Municipal Incinerator," *J. Air Poll. Control Assn.,* **22,** 248 (1972).
103. "Can Plastics be Incinerated Safely?," *Environ. Sci. Technol.,* **5,** 667 (1971).
104. "Man-Made Fiber Fact Book," published annually by the Man-Made Fiber Producers Assoc., Inc., 100 Connecticut Ave., Washington, D.C. 10036.
105. Kaiser, E. R. "The Sulfur Balance of Incinerators," *J. Air Poll. Control Assn.,* **18,** 171 (1968).
106. Stephens, G. R., L. Hankin, and W. D. Glover. "Note on Lead Emissions From Incinerated Sewage Sludge Detected on Tree Foliage," *J. Air Poll. Control Assn.,* **22,** 799 (1972).
107. Peterson, M. L., and F. J. Stutzenberger. "Microbiological Evaluation of Incinerator Operations," *App. Microbiol.,* **18,** 8 (1969).

108. Heaney, F. L. "Air Pollution Controls at Braintree Incinerator," *J. Air Poll. Control Assn.*, **22**, 617 (1972).

109. Zausner, E. R. "An Accounting System for Incinerator Operations," U.S. Department of Health, Education and Welfare, Public Health Service Publication No. 2032 (1970).

110. Feldstein, M. "Studies on the Analysis of Hydrocarbons from Incinerator Effluents with a Flame Ionization Detector," *J. Air Poll. Control Assn.*, **12**, 139 (1962).

111. Rehm, F. R. "Test Methods for Determining Emission Characteristics of Incinerators," *J. Air Poll. Control Assn.*, **15**, 127 (1965).

112. Achinger, W. C., and L. E. Daniels. "An Evaluation of Seven Incinerators," presented at the 1970 National Incinerator Conference, EPA publication, SW 51ts.lj, Solid Waste Management Series.

113. Feuss, J. V., and F. B. Flower. "State of the Art: The Design of Apartment House Incinerators," *J. Air Poll. Control Assn.*, **19**, 142 (1969).

114. Sableski, J. J., and W. A. Cote. "Air Pollutant Emissions from Apartment House Incinerators," *J. Air Poll. Control Assn.*, **22**, 239 (1972).

115. Gerstle, R. W., and D. A. Kemnitz. "Atmospheric Emissions from Open Burning," *J. Air Poll. Control Assn.*, **17**, 324 (1967).

116. Kurker, C. "Reducing Emissions from Refuse Disposal," *J. Air Poll. Control Assn.*, **19**, 69 (1969).

117. Hamburg, F. C. "Economically Feasible Alternatives to Open Burning in Railroad Freight Car Dismantling," *J. Air Poll. Control Assn.*, **19**, 477 (1969).

118. "Disposal of Bulky Solid Wastes," *J. Air Poll. Control Assn.*, **22**, 858 (1972).

119. Boubel, R. W., E. F. Darley, and E. A. Schuck. "Emissions from Burning Grass Struble and Straw," *J. Air Poll. Control Assn.*, **19**, 497 (1969).

120. Paulus, H. J. "Nuisance Abatement by Combustion," in *Air Pollution, Vol. III, Sources of Air Pollution and Their Control*, A. C. Stern, Ed. (New York: Academic Press, 1968) chapter 48.

121. "Air Pollution Aspects of Tepee Burners," U.S. Dept. of Health, Education and Welfare Publication AP-028 (Sept., 1966), Available NTIS PB 173, 986.

122. Coleman, L. W., and L. F. Cheek. "Liquid Waste Incineration," *Chem. Eng. Prog.*, **64**, 83 (1968).

123. Santoleri, J. J. "Chlorinated Hydrocarbon Waste Disposal and Recovery Systems," *Chem. Eng. Prog.*, **69**, 68 (1973).

124. Miller, M. R., and H. J. Wilhoyte. "A Study of Catalyst Support Systems for Fume Abatement of Hydrocarbon Solvents," *J. Air Poll. Control Assn.*, **17**, 791 (1967).

125. Werner, K. D. "Catalytic Oxidation of Industrial Waste Gas," *Chem. Eng.* (November 4, 1968) p 179.

126. Essenhigh, R. H. "Incineration—A Practical and Scientific Approach," *Environ. Sci. Technol.*, **2**, 525 (1968).

127. Rogers, J. E. L., A. F. Sarofim, J. B. Howard, G. C. Williams, and D. H. Fine. "Combustion Characteristics of a Simulated Refuse," presented at the 1973 Technical Session of the Central States Section of the Combustion Institute, Champaign, Ill. (March 1973).

128. "Fluidized Bed Incinerators Studied for Solid Waste Disposal," *Environ. Sci. Technol.*, **2**, 495 (1968).

129. Pradt, L. A. "Developments in Wet Air Oxidation," *Chem. Eng. Prog.*, **68**, 72 (1972).

130. Shuster, W. W. "Partial Oxidation of Solid Organic Wastes," U.S. Dept. of Health, Education and Welfare Publication #2133 (1970).

131. Hoffman, D. A., and R. A. Fritz. "Batch Retort Pyrolysis of Solid Municipal Wastes," *Environ. Sci. Technol.*, **2**, 1023 (1968).

132. "Pyrolysis of Refuse Gains Ground," *Environ. Sci. Technol.*, **5**, 310 (1971).

133. "Compilation of Air Pollution Emission Factors," U.S. Environmental Protection Agency, AP-42, April, 1973, 2nd Edition.

134. "Air Pollution Aspects of Emission Sources: Iron and Steel Mills—A Bibliography with Abstracts," U.S. Environmental Protection Agency Publication AP-107, May, 1972.

135. Sebesta, W. "Ferrous Metallurgical Processes," in *Air Pollution, Vol. III, Sources of Air Pollution and Their Control,* A. C. Stern, Ed. (New York: Academic Press, 1968) chapter 36.

136. Nelson, K. W. "Nonferrous Metallurgical Operations," in *Air Pollution, Vol. III, Sources of Air Pollution and Their Control,* A. C. Stern, Ed. (New York: Academic Press, 1968) chapter 37.

137. "Air Pollution Aspects of Emission Sources: Primary Lead Production—A Bibliography with Abstracts," Environmental Protection Agency Publication AP-126, June, 1973.

138. "Air Pollution Aspects of Emission Sources: Primary Copper Production—A Bibliography with Abstracts," U.S. Environmental Protection Agency Publication AP-125, June, 1973.

139. "Air Pollution Aspects of Emission Sources: Primary Aluminum Production—A Bibliography with Abstracts," U.S. Environmental Protection Agency Publication AP-119, June, 1973.

140. Sussman, V. H. "Nonmetallic Mineral Products Industries," in *Air Pollution, Vol. III, Sources of Air Pollution and Their Control,* A. C. Stern, Ed. (New York: Academic Press, 1968) chapter 35.

141. "Air Pollution Aspects of Emission Sources: Petroleum Refineries—A Bibliography with Abstracts," U.S. Environmental Protection Agency Publication AP-110, June, 1972.

142. Elkin, H. F. "Petroleum Refinery Emissions," in *Air Pollution, Vol. III, Sources of Air Pollution and Their Control,* A. C. Stern, Ed. (New York: Academic Press, 1968) chapter 34.

143. Heller, A. N., S. T. Cuffe, and D. R. Goodwin. "Inorganic Chemical Industry," in *Air Pollution, Vol. III, Sources of Air Pollution and Their Control,* A. C. Stern, Ed. (New York: Academic Press, 1968) chapter 38.

144. Adams, D. F. "Pulp and Paper Industry," in *Air Pollution, Vol. III, Sources of Air Pollution and Their Control,* A. C. Stern, Ed. (New York: Academic Press, 1968) chapter 39.

145. Faith, W. L. "Food and Feed Industries," in *Air Pollution, Vol. III, Sources of Air Pollution and Their Control,* A. C. Stern, Ed. (New York: Academic Press, 1968) chapter 40.

146. Lichty, L. C. *Internal Combustion Engines,* 6th ed. (New York: McGraw-Hill, 1951).

147. Obert, E. F. *Internal Combustion Engines,* 3rd ed. (Scranton, Pa.: International Textbook Co., 1968).

148. *Engine Emissions, Pollutant Formation and Measurement,* G. S. Springer and D. J. Patterson, Eds. (New York: Plenum Press, 1973).

149. Patterson, D. J., and N. A. Henein. *Emissions from Combustion Engines and Their Control* (Ann Arbor, Michigan: Ann Arbor Science Publishers, 1972).

150. *Emissions from Continuous Combustion Systems,* W. Cornelius and W. G. Agnew, Eds. (New York: Plenum Press, 1972).

151. "Vehicle Emissions—Part I," *SAE Progress in Technology Series, Volume 6* (New York: Society of Automotive Engineers, 1964).

152. "Vehicle Emissions—Part II," *SAE Progress in Technology Series, Volume 12* (New York: Society of Automotive Engineers, 1967).

153. "Vehicle Emissions—Part III," *SAE Progress in Technology Series, Volume 14* (New York: Society of Automotive Engineers, 1971).

154. "The Automobile and Air Pollution: A Program for Progress," a report by the Panel on Electrically Powered Vehicles, R. S. Morse, Chairman to the U.S. Department of Commerce Technical Advisory Board, October, 1967.

155. "Semiannual Report by the Committee on Motor Vehicle Emissions of the National Academy of Sciences to the U.S. Environmental Protection Agency," National Academy of Sciences, Washington, D.C., January, 1972.

156. "Report by the Committee on Motor Vehicle Emissions of the National Academy of Sciences to the U.S. Environmental Protection Agency," National Academy of Sciences, Washington, D.C., February, 1973.

157. "Automobile Emission Control—The State of the Art as of December, 1972," a report to the Administrator, Environmental Protection Agency. Prepared by the Division of Emission Control Technology, Mobil Source Pollution Control Program, Office of Air and Water Programs, U.S. Environmental Protection Agency, December, 1972.

158. "Cumulative Regulatory Effects on the Cost of Automotive Transportation (RECAT)," Final Report of the Ad Hoc Committee, prepared for the Office of Science and Technology, February, 1972.

159. "The Impact of Auto Emission Standards," Report of the Staff of the Subcommittee on Air and Water Pollution to the Committee on Public Works, United States Senate. Serial No. 93-11, October, 1973.

160. Hearings on the subject of mobile source emissions are held by various legislative groups. Two of them are the Subcommittee on Air and Water Pollution, Committee on Public Works (U.S. Senate) and the Public Health and Environment Subcommittee of the Interstate and Foreign Commerce Committee (U.S. House of Representatives).

161. Kruse, R. E., and T. A. Huls. "Development of the Federal Urban Driving Schedule," presented at the Automobile Engineering Meeting, Society of Automobile Engineers, Detroit, Michigan, May 1973, Paper No. 730553.

162. Huls, T. A. "Evolution of Federal Light-Duty Mass Emission Regulations," presented at the Automobile Engineering Meeting, Society of Automobile Engineers, Detroit, Michigan, May 1973, Paper No. 730554.

163. "Exhaust Emissions from Privately Owned 1966-1972 California Automobiles. A Statistical Evolution of Surveillance Data," reports issued periodically by the Staff of the California Air Resources Board.

164. "A Study of Mandatory Engine Maintenance for Reducing Exhaust Emissions, Volume I, Executive Summary," prepared in support of APRAC project CAPE 13-68 for the Coordinating Research Council, Inc. and the U.S. Environmental Protection Agency by TRW Transportation and Environmental Operations & Scott Research Laboratories, Inc., July, 1973.

165. Fegraus, C. E., C. J. Domke, and J. Marzen. "Contribution of the Vehicle Population to Atmospheric Pollution," presented at the SAE Automobile Engineering Meeting, Detroit, Michigan, May, 1973, Paper No. 730530.

166. Springer, K. J., and C. D. Tyree. "Exhaust Emissions from Gasoline-Powered Vehicles Above 6000 lb. Gross Vehicle Weight," report prepared for U.S. Environmental Protection Agency by Southwest Research Institute, San Antonio, Texas, April, 1972.

167. Hare, C. T., and K. J. Springer. "Exhaust Emissions from Uncontrolled Vehicles and Related Equipment using Internal Combustion Engines, Part 3—Motorcycles," report prepared for the U.S. Environmental Protection Agency by the Southwest Research Institute, San Antonio, Texas, March, 1973.

168. Hare, C. T., and K. J. Springer. "Small Engine Emissions and the Impact," Automotive Engineering, 80, 15 (1972).

169. Hare, C. T., and K. J. Springer. "Exhaust Emissions from Uncontrolled Vehicles and Related Equipment Using Internal Combustion Engines, Part I—Locomotive Diesel Engines and Marine Counterparts," prepared for the U.S. Environmental Protection Agency by Southwest Research Institute, San Antonio, Texas, October, 1972.

170. George, R. E., J. S. Nevitt, and J. A. Verssen. "Jet Aircraft Operations: Impact on the Air Environment," J. Air. Poll. Control Assn., 22, 451 (1972).

171. Sallee, G. P. "Status Report on Aircraft and Airports as Sources of Pollution," presented at the Conference on Aircraft and the Environment—Part 2, conducted jointly by the Society of Automotive Engineers and U.S. Department of Transportation, Society of Automotive Engineers Paper No. 710322.

172. "Automobile Fuel Economy," available from Motor Vehicle Manufacturers Association of the United States, Inc., 320 New Center Building, Detroit, Michigan 48292, 1973.

173. Eccleston, B. H., B. F. Noble, and R. W. Hurn. "Influence of Volatile Fuel Components on Vehicle Emissions," U.S. Bureau of Mines Report of Investigation No. 7291 (1970). Also, see B. H. Eccleston, and R. W. Hurn, "Effect of Fuel Front-End and Midrange Volatility on Automobile Emissions," U.S. Bureau of Mines Report of Investigation No. 7707 (1972).

174. Clarke, P. J. "The Effect of Gasoline Volatility on Exhaust Emissions," presented at the SAE National Fuels and Lubricants Meeting, Tulsa, Oklahoma, November 1972. Paper No. 720932.

175. Bond, W. D. "Quick Heat Intake Manifolds for Reducing Cold Engine Emissions," presented at the SAE National Fuels and Lubricants Meeting, Tulsa, Oklahoma, November 1972, Paper No. 72935.

176. "Automotive Spark Ignition Engine Emission Control Systems to Meet the Requirements of the 1970 Clean Air Act Amendments," Report of the Emission Control Systems Panel to the Committee on Motor Vehicle Emissions, National Academy of Sciences, May, 1973.

177. Rivard, J. G. "Closed Loop Electronic Fuel Injection Control of the Internal Combustion Engine," presented at the SAE International Automotive Engineering Congress, Detroit, Michigan, January 1973, Paper No. 730005.

178. Edwards, J. B., and D. M. Teague. "Unraveling the Chemical Phenomena Occurring in Spark Ignition Engines," presented at the Society of Automotive Engineers Mid-Year Meeting, Detroit, Michigan, May 1970, Paper No. 700489.

179. Edward, J. B. "The Chemistry of Spark-Ignition Engine Combustion and Emission Formation," in *Engine Emissions-Pollutant Formation and Measurement*, G. S. Springer and D. J. Patterson, Eds. (New York: Plenum Press, 1973) chapter 2.

180. Daniel, W. A. "Why Engine Variables Affect Exhaust Hydrocarbon Emissions," presented at the Society of Automotive Engineers Automotive Engineering Congress, Detroit, Michigan, January 1970, Paper No. 700108.

181. Wentworth, J. T. "Effect of Combustion Chamber Surface Temperature on Exhaust Hydrocarbon Concentration," presented at the SAE Mid-Year Meeting, Montreal, Quebec, Canada, June 1971, Paper No. 710587.

182. Haskell, W. W., and C. E. Legate. "Exhaust Hydrocarbon Emissions from Gasoline Engines—Surface Phenomena," presented at the SAE Automotive Engineering Congress, January 1972, Paper No. 720255.

183. Komiyama, K., and J. B. Heywood. "Predicting NO_x Emissions and Effects of Exhaust Gas Recirculation in Spark-Ignition Engines," presented at the SAE Automotive Engineering Meeting, Detroit, Michigan, May 1973, Paper No. 730475.

184. Lavoie, G. A., J. B. Heywood, and J. C. Keck. "Experimental and Theoretical Study of Nitric Oxide Formation in Internal Combustion Engines," *Comb. Sci. Technol.*, 1, 313 (1970).

185. Starkman, E. S., H. E. Steward, and V. A. Zvonow. "An Investigation into the Formation and Modification of Emission Precursors," presented at the SAE International Automotive Engineering Congress, Detroit, Michigan, January 1969, Paper No. 690020.

186. Newhall, H. K. "Kinetics of Engine Generated Nitrogen Oxides and Carbon Monoxide," *Proc. 12th Symp. (International) on Combustion*, Poitiers, France, July 1968 (Pittsburgh, Pa.: Combustion Institute, 1969).

187. Blumberg, P., and J. T. Kummer. "Prediction of NO Formation in Spark-

Ignited Engines—An Analysis of Methods of Control," *Comb. Sci. Technol.*, 4, 73 (1971).

188. Eyzat, P., and J. C. Guibet. "A New Look at Nitrogen Oxide Formation in Internal Combustion Engines," presented to the SAE Automobile Engineering Congress, Detroit, Michigan, January 1968, Paper No. 680124.

189. Aiman, W. R. "Engine Speed and Load Effects on Charge Dilution and Nitric Oxide Emission," presented at the SAE Automobile Engineering Congress, Detroit, Michigan, January 1972, Paper No. 720256.

190. Brehob, W. M. "Mechanisms of Pollutant Formation and Control from Automotive Sources," SAE publication SP-365, *Engineering Know-How in Engine Design—Part 19* (New York: Society of Automotive Engineers, 1971).

191. Lestz, S. S., W. E. Meyer, and C. R. Colony. "Emission from a Direct Cylinder Water-Injected Spark-Ignition Engine," presented at the SAE Automobile Engineering Congress, Detroit, Michigan, January, 1972 Paper No. 720113.

192. Nicholls, J. E., I. A. El-Messire, and H. K. Newhall. "Inlet Manifold Water Injection for Control of Nitrogen Oxides—Theory and Experiment," presented at the SAE International Automobile Engineering Congress, Detroit, Michigan, January, 1969, Paper No. 690018.

193. Hansel, J. G. "Lean Automobile Engine Operation—Hydrocarbon Exhaust Emissions and Combustion Characteristics," presented at the SAE Automobile Engineering Congress, Detroit, Michigan, January, 1971, Paper No. 710164.

194. Barton, R. K., S. S. Lestz, and W. E. Meyer. "An Empirical Model for Correlating Cycle by Cycle Cylinder Gas Motion and Combustion Variations of a Spark-Ignition Engine," presented at the SAE Automobile Engineering Congress, Detroit, Michigan, January 1971, Paper No. 710163.

195. Matsuoka, S., T. Yamagucki, and Y. Umemura. "Factors Influencing the Cyclic Variation of Combustion of Spark Ignition Engines," presented at the SAE Mid-Year Meeting, Montreal, Quebec, Canada, June 1971, Paper No. 710586.

196. Anderson, W. J., S. S. Lestz, and W. E. Meyer. "The Effect of Charge Dilution on CBC Variations and Exhaust Emissions of an SI Engine," presented at the SAE International Automobile Engineering Congress, Detroit, Michigan, January 1973, Paper No. 730152.

197. Sorenson, S. C., P. S. Myers, and O. A. Uyehara. "The Reaction of Ethane in Spark Ignition Engine Exhaust Gas," presented at the SAE Mid-Year Meeting, Detroit, Michigan, May 1970, Paper No. 700471.

198. Sigworth, H. W., P. S. Myers, and O. A. Uyehara. "The Disappearance of Ethylene, Propylene, n-Butane and 1-Butane in Spark-Ignition Engine Exhaust," presented at the SAE Mid-Year Meeting, Detroit, Michigan, May 1970, Paper No. 700472.

199. Patterson, D. J., R. H. Kadlec, B. Carnahan, H. A. Lord, J. J. Martin, W. Mirsky, and E. Sondreal. "Kinetics of Oxidation and Quenching of Combustibles in Exhaust Systems of Gasoline Engines," Annual Progress Report No. 3 (Final Report). Prepared for the Coordinating Research Council and the U.S. Environmental Protection Agency. APRAC Project CAPE 8-68, 1972.

200. Cantwell, E. N., R. A. Hoffman, I. T. Roselund, and S. W. Ross. "A Systems Approach to Vehicle Emissions Control," presented to the SAE National Automobile Engineering Meeting, Detroit, Michigan, May 1972, Paper No. 720510.

201. Lang, R. J. "A Well-Mixed Thermal Reactor System for Automotive Emission Control," presented at the SAE Mid-Year Meeting, Montreal, Quebec, Canada, June 1971, Paper No. 710608.

202. Jaimee, A., D. E. Schneider, A. I. Rozmanith, and J. W. Sjoberg. "Thermal Reactor—Design, Development and Performance," presented to the SAE Automotive Engineering Congress, Detroit, Michigan, January 1971, Paper No. 710151.

203. Pozniak, D. J., and R. M. Siewert. "Continuous Secondary Air Modulation—Its Effects on Thermal Manifold Reactor Performance," presented at the SAE Automobile Engineering Meeting, Detroit, Michigan, May 1973, Paper No. 730493.

204. Schwing, R. C. "An Analytical Framework for the Study of Exhaust Manifold Reactor Oxidation," presented at the SAE Automotive Engineering Congress, Detroit, Michigan, January 1970, Paper No. 700109.

205. Jagel, K. I., and F. G. Dwyer. "HC/CO Oxidation Catalysts for Vehicle Exhaust Emission Control," presented at the SAE Automotive Engineering Congress, Detroit, Michigan, January 1971, Paper No. 710290.

206. Hancock, E. E., R. M. Campau, and R. Connolly. "Catalytic Converter Vehicle System Performance: Rapid *vs.* Customer Mileage," presented at the SAE Automotive Engineering Congress, Detroit, Michigan, January 1971, Paper No. 710292.

207. Roth, J. F., and J. W. Gambell. "Control of Automotive Emissions by Particulate Catalysts," presented at the SAE International Automotive Engineering Congress, Detroit, Michigan, January 1973, Paper No. 730277.

208. "Aldehyde and Reactive Emissions from Motor Vehicles. Part I—Aldehyde and Reactive Emissions from Advanced Automotive Control System Vehicles. Part II—Characterization of Motor Vehicle Aldehyde and Reactive Organic Emissions from 1970 through 1973 Model Vehicles," Bartlesville Energy Research Center, U.S. Bureau of Mines, March 1973.

209. Neal, A. H., E. E. Wigg, and E. L. Holt. "Fuel Effects on Oxidation Catalysts and Catalyst Equipped Vehicles," presented at the SAE Mid-Year Meeting, Detroit, Michigan, May 1973, Paper No. 730593.

210. Malbin, M. D., A. F. Soby, and W. E. Haskell. "On-line Detection of Polynuclear Aromatic Hydrocarbons in Automobile Exhaust I. Evaluation of Catalytic Converters," presented at the 66th Annual Meeting of the Air Pollution Control Assn., Chicago, Ill., June 1973, Paper No. 73-77.

211. Gross, G. P. "The Effect of Fuel and Vehicle Variables on Polynuclear Aromatic Hydrocarbon and Phenol Emissions," presented at the SAE Automotive Engineering Congress, Detroit, Michigan, January 1972, Paper No. 720210.

212. Meguerian, G. H., and C. R. Lang. "NO_x Reduction Catalysts for Vehicle Emissions Control," presented at the SAE Automotive Engineering Congress, Detroit, Michigan, January 1971, Paper No. 710291.

213. Lunt, R. S., L. S. Bernstein, J. G. Hansel, and E. L. Holt. "Application of a Monel-Platinum Dual-Catalyst System to Automotive Emission Control," presented to the SAE Automotive Engineering Congress, Detroit, Michigan, January 1972, Paper No. 720209.

214. Retzloff, J. B., L. Plonsker, and R. B. Sneed. "Fuel Detergency—Effect on Emissions," presented at the SAE National Fuels and Lubricants Meeting, Tulsa, Oklahoma, November 1972, Paper No. 720941.

215. Mixom, L. W., A. I. Rozmanith, and W. T. Wotring. "Effect of Fuel and Lubricant Additives on Exhaust Emissions," presented at the SAE Automotive Engineering Congress, Detroit, Michigan, January 1971, Paper No. 710295.

216. Moran, J. B., O. J. Manary, R. H. Fay, and M. J. Baldwin. "Development of Particulate Emission Control Technique for Spark Ignition Engines," prepared for the U.S. Environmental Protection Agency, Office of Air Programs, Ann Arbor, Michigan, by the Dow Chemical Corp., Midland, Michigan, July 1971.

217. Fleming, R. D., J. R. Allsup, T. R. French, and D. E. Eccleston. "Propane as an Engine Fuel for Clean Air Requirements," *J. Air Poll. Control Assn.*, **22**, 451 (1972).

218. Genslak, S. L. "Evaluation of Gaseous Fuels for Automobiles," presented to the SAE Automotive Engineering Congress, Detroit, Michigan, January 1972, Paper No. 720125.

219. Malte, P. C., and G. A. Wooldridge. "Low Pollution Characteristics of Urban Transient Buses Fueled with Liquified Natural Gas," presented at the SAE National West Coast Meeting, San Francisco, California, August 1972, Paper No. 720685.
220. "Alcohols and Hydrocarbons as Motor Fuels," SAE publication SP-254 (New York: Society of Automotive Engineers, 1964).
221. Ebersole, G. D., and F. S. Manning. "Engine Performance and Exhaust Emissions: Methanol vs. Isooctane," presented at the SAE National West Coast Meeting, San Francisco, California, August 1972, Paper No. 720692.
222. Adelman, H. G., D. G. Andrews, and R. S. Devoto. "Exhaust Emissions from a Methanol Fueled Automobile," presented at the SAE National West Coast Meeting, San Francisco, California, August 1972, Paper No. 720693.
223. Ninomiya, J. S., and A. Golovoy. "Effects of Air Fuel Ratio on Composition of Hydrocarbon Exhaust from Isooctane, Diisobutylene, Toluene and Toluene-n-Heptane Mixture," presented at the SAE Mid-Year Meeting, Chicago, Illinois, May 1969, Paper No. 690504.
224. Sawyer, R. F., E. S. Starkman, L. Muzio, and W. L. Schmidt. "Oxides of Nitrogen in the Combustion Products of an Ammonia Fueled Reciprocating Engine," presented at the SAE Mid-Year Meeting, Detroit, Mich., May 1968, Paper No. 680401.
225. Ninomiya, J. S., W. Bergman, and B. H. Simpson. "Automotive Particulate Emissions," presented at the 2nd International Clean Air Congress of the International Union of Air Pollution Prevention Association, Washington, D.C., December 1970, Paper No. EN-10G.
226. Habibi, K. "Characterization of Particulate Lead in Vehicle Exhausts—Experimental Results," *Environ. Sci. Technol.* 4, 239 (1970).
227. Schuchmann, H. P., and K. J. Laidler. "Nitrogen Compounds other than NO in Automobile Exhaust Gas," *J. Air Poll. Control Assn.*, 22, 52 (1972).
228. Harkins, J. II., and S. W. Nicksic. "Ammonia in Auto Exhaust," *Environ. Sci. Technol.*, 1, 751 (1967).
229. Begeman, C. R., and J. M. Colucci. "Polynuclear Aromatic Hydrocarbon Emissions from Automotive Engines," presented at the SAE Mid-Year Meeting, Detroit, Michigan, May 1970, Paper No. 700469.
230. Seizinger, D. E., and B. Dimitriades. "Oxygenates in Automotive Exhaust Gas-Estimation of the Levels of Carbonyls and Non-carbonyls in Exhaust from Gasoline Fuels," U.S. Bureau of Mines Report of Investigation No. 7675, 1972.
231. Seizinger, D. E., and B. Dimitriades. "Oxygenates in Exhaust from Simple Hydrocarbon Fuels," *J. Air Poll. Control Assn.*, 22, 47 (1972).
232. Wigg, E. E., R. J. Campion, and W. L. Peterson. "The Effect of Fuel Hydrocarbon Composition on Exhaust Hydrocarbon and Oxygenate Emissions," presented to the SAE Automotive Engineering Congress, Detroit, Michigan, January 1972, Paper No. 720251.
233. Tataiak, K. "Exhaust Emission Characteristics of a Mercedes 200 D. Diesel Engine Car," presented at the Central States Section Meeting of the Combustion Institute, Champaign, Ill., March 1973.
234. Brisson, B., A. Ecomard, and P. Eyzat. "A New Diesel Combustion Chamber—The Variable Throat Chamber," presented at the SAE International Automotive Engineering Congress, Detroit, Michigan, January 1973, Paper No. 730167.
235. Khan, I. M., C. H. T. Wang, and B. E. Langridge. "Effect of Air Swirl on Smoke and Gaseous Emissions from Direct-Injection Diesel Engines," presented at the SAE Automotive Engineering Congress, Detroit, Michigan, January 1972, Paper No. 720102.
236. Hames, R. J., D. F. Merrion, and H. S. Ford. "Some Effects of Fuel Injection

System Parameters on Diesel Exhaust Emissions," presented at the SAE National West Coast Meeting, Vancouver, B.C., Canada, August 1971, Paper No. 710671.

237. Bosecker, R. E., and D. F. Webster. "Precombustion Chamber Diesel Engine Emissions—A Progress Report," presented at the SAE National West Coast Meeting, Vancouver, B.C., Canada, August 1971, Paper No. 710672.

238. Tataiak, K. "Exhaust Emission Characteristics of a Mercedes 200D Diesel Engine Car," presented at the Central States Section Meeting of the Combustion Institute, Champaign, Ill., March 1973.

239. Henein, N. A. "Diesel Engine Combustion and Emissions" in *Engine Emissions—Pollutant Formation and Measurement,* G. S. Springer and D. J. Patterson, Eds. (New York: Plenum Press, 1973) chapter 6.

240. Benson, R. S. "A Comprehensive Digital Computer Program to Simulate a Compression Ignition Engine Including Intake and Exhaust Systems," presented to the SAE Automotive Engineering Congress, Detroit, Michigan, January 1971, Paper No. 710173.

241. Frey, J. W., and M. Corn. "Physical and Chemical Characteristics of Particles in Diesel Exhaust," *Amer. Ind. Hygiene Assn. J.,* **28**, 468 (1967).

242. Ray, S. K., and R. Long. "Polycyclic Aromatic Hydrocarbons from Diffusion Flames and Diesel Engine Combustion," *Combust. Flame,* **6**, 139 (1964).

243. Turley, C. D., D. L. Brenchley, and R. R. Landott. "Barium Additives as Diesel Smoke Suppressants," *J. Air Poll. Control Assn.,* **23**, 783 (1973).

244. Aaronson, A. E., and R. A. Matula. "Diesel Odor and the Formation of Aromatic Hydrocarbons during the Heterogenous Combustion of Pure Cetane in a Single-Cylinder Diesel Engine," *Proc. 13th Symp. (International) on Combustion,* Salt Lake City, Utah, (Pittsburgh, Pa.: The Combustion Institute, 1971) p 471.

245. "Chemical Identification of the Odor Components in Diesel Engine Exhaust," final report to the Coordinating Research Council and NAPCA, U.S. Public Health Service, Arthur D. Little, Inc., July 1969.

246. "Analysis of the Odorous Compounds in Diesel Engine Exhaust" final report to the Coordinating Research Council and Environmental Protection Agency, Arthur D. Little, Inc., June 1972.

247. Stahman, R. C., G. D. Kittredge, and K. J. Springer. "Smoke and Odor Control for Diesel-Powered Trucks and Buses," presented at the SAE Mid-Year Meeting, Detroit, Michigan, May 1968, Paper No. 680443.

248. Colucci, J. M., and G. J. Barnes. "Evaluation of Vehicle Exhaust Gas Odor Intensity Using Natural Dilution" presented to the SAE Automobile Engineering Congress, Detroit, Michigan, January 1970, Paper No. 700105.

249. Turk, A. "Selection and Training of Judges for Sensory Evaluation of the Intensity and Character of Diesel Exhaust Odors," U.S. Department of Health, Education and Welfare, 1967.

250. Springer, K. J., and H. E. Dietzmann. "Diesel Exhaust Hydrocarbon Measurement—A Flame Ionization Method," presented at the SAE Automotive Engineering Congress, Detroit, Michigan, January 1970, Paper No. 700106.

251. Smith, D. S., R. F. Sawyer, and E. S. Starkman. "Oxides of Nitrogen from Gas Turbines," *J. Air Poll. Control Assn.,* **18**, 30 (1968).

252. Lozano, E. R., W. W. Melvin, and S. Hochheiser. "Air Pollution Emissions from Jet Engines," *J. Air Poll. Control Assn.,* **18**, 392 (1968).

253. Williamson, R. C., and J. A. Russell. "On-Line Gas Analysis of Jet Engine Exhaust," presented at the SAE Combined Fuels and Lubricants, Powerplant and Transportation Meetings, Pittsburgh, Pa., November 1967, Paper No. 670945.

254. Tuttle, J. H., R. A. Altenkirch, and A. M. Mellor. "Emissions from and Within an Allison J-33 Combustor, II: The Effect of Inlet Air Temperature," presented at the 1973 Central States Section Meeting of the Combustion Institute, Champaign, Ill., March 1973.

255. Gross-Gronomski, L. "Smoke in Gas Turbine Exhaust," presented at the ASME Winter Annual Meeting and Energy Systems Exposition, Pittsburgh, Pa., November 1967, Paper No. 67-WA/GT-5.

256. Haupt, C. G. "Exhaust Emission by a Small Gas Turbine," presented at the SAE Mid-Year Meeting, Detroit, Mich., May 1968, Paper No. 680463.

257. Lieberman, A. "Composition of Exhaust from a Regenerative Turbine System," *J. Air Poll. Control Assn.*, **18**, 149 (1968).

258. Cornelius, W., D. L. Stivender, and R. E. Sullivan. "A Continuous Combustion System for a Vehicular Regenerative Gas Turbine Featuring Low Air Pollutant Emissions," presented at the SAE Combined Fuels and Lubricants, Powerplant and Transportation Meetings, Pittsburgh, Pa., November 1967, Paper No. 670936.

259. Saintsbury, J. A., P. Sampath, and H. G. Eatock. "Simple Automobile Gas Turbine Combustors for Low Emissions," presented at the SAE Combined Commercial Vehicle Engineering & Operations and Powerplant Meetings, Chicago, Ill., June 1973, Paper No. 730670.

260. Korth, M. W., and A. H. Rose, Jr. "Emissions from a Gas Turbine Automobile," presented at the SAE Mid-Year Meeting, Detroit, Mich., May 1968, Paper No. 680402.

261. Norgren, C. T., and R. D. Ingebo. "Effects of Prevaporized Fuel on Exhaust Emissions from an Experimental Gas Turbine Combustor," presented at the Spring Meeting of the Central States Section of the Combustion Institute, Champaign, Ill., March 1973.

262. Gleason, J. G., and J. J. Faitani. "Smoke Abatement in Gas Turbine Engines Through Combustor Design," presented at the SAE Mid-Continent Section Meeting, February 1967, Paper No. 670200.

263. Hussmann, A. W., and G. W. Maybach. "The Film Vaporization Combustor," *SAE Trans.*, **69**, 563 (1961).

264. LaPointe, C. W., and W. L. Schultz. "Comparison of Emission Indices within a Turbine Combustor Operated on Diesel Fuel or Methanol," presented at the SAE National Powerplant Meeting, Chicago, Ill., June 1973, Paper No. 730669.

265. Chinitz, W. "Rotary Engines," *Scientific American*, **220**, 90 (1969).

266. Froede, W. G. "NSU's Double Bank Production Rotary Engine," presented at the SAE Mid-Year Meeting, Detroit, Mich., May 1968, Paper No. 680461.

267. Yamamoto, K., T. Muroki, and T. Kobayakawa. "Combustion Characteristics of Rotary Engines," presented at the SAE Metropolitan Section Meeting, April 1972, Paper No. 720357.

268. Bracco, F. V. "Theoretical Analysis of Stratified, Two-Phase Wankel Engine Combustion," presented at the Spring Meeting of the Central States Section of the Combustion Institute, Champaign, Ill., March 1973.

269. Simko, A., M. A. Choma, and L. L. Repko. "Exhaust Emission Control by the Ford Programed Combustion Process—PROCO," presented at the SAE Automotive Engineering Congress, Detroit, Mich., January 1972, Paper No. 720052.

270. Miyake, M. "Developing a New Stratified Charge Combustion System with Fuel Injection for Reducing Exhaust Emissions on Small Farm and Industrial Engines," presented at the SAE Automotive Engineering Congress, January 1972, Paper No. 720196.

271. Bascunana, J. L. "Divided Combustion Chamber Gasoline Engines—A Review for Emissions and Efficiency," presented at the 66th Annual Meeting, Air Pollution Control Association, Chicago, Ill., June 1973, Paper No. 73-74.

272. "Technical Report on Honda CVCC System," prepared for the Subcommittee on Public Works, U.S. Senate, by Honda Motor Co., Japan, May 17, 1973.

273. Mitchell, E., J. M. Cobb, and R. A. Frost. "Design and Evaluation of a Stratified Charge Multifuel Military Engine," presented at the SAE Automotive Engineering Congress, Detroit, Mich., January 1968, Paper No. 680042.

274. Blumberg, P. N. "Nitric Oxide Emissions from Stratified Charge Engines: Prediction and Control," presented at the 1973 Central States Section Meeting of the Combustion Institute, Champaign, Ill., March 1973.

275. Bellan, J. R., and W. A. Sirignano. "A Theory of Turbulent Flame Development in Stratified Charge Internal Combustion Engines," presented at the 1973 Central States Section Meeting of the Combustion Institute, Champaign, Ill., March 1973.

276. Welsh, H. W. "Study of Low Emission Vehicle Power Plants Using Gaseous Working Fluids," prepared for the U.S. Environmental Protection Agency, Office of Air Programs, Advanced Automotive Power Systems Division by Thermo Mechanical Systems Co., Canoga Park, Cal., August 1972. Available from NTIS PB-220 148.

277. Gerstman, J., and F. Pompei. "Performance of a Homogeneous Combustor for a Rankine Cycle Steam Engine," presented at the SAE Combined Commercial Vehicle Engineering & Powerplant Meetings, Chicago, Ill., June 1973, Paper No. 730671.

278. Doerner, W. A., R. J. Dietz, O. R. Van Buskirk, S. B. Levy, P. J. Rennolds, and M. F. Bechtold. "A Rankine Cycle Engine with Rotary Heat Exchangers," presented at the SAE Automotive Engineering Congress, Detroit, Mich., January 1972, Paper No. 720053.

279. Sakhuja, R. K., and A. D. Vasilakis. "Low Emission Combustor Development for Automotive Rankine-Cycle Engines," presented at the SAE Combined Commercial Vehicle Engineering & Powerplant Meetings, Chicago, Ill., June 1973, Paper No. 730672.

280. Compton, W. A., J. R. Shekleton, T. E. Duffy, and R. T. LeCren. "Low Emissions from Controlled Combustion for Automotive Rankine Cycle Engines," *J. Air Poll. Control Assn.*, **22**, 699 (1972).

281. Huffman, G. D., and O. E. Buxton. "Thermodynamic Optimization of Rankine Cycle Space Power Systems," *J. Engrg. Power,* **90**, 89 (1968).

282. Bjerklie, J. W., and B. Sternlicht. "Critical Comparison of Low Emission Otto and Rankine Engines for Automotive Use," presented at the International Automotive Engineering Congress, Detroit, Mich., January 1969, Paper No. 690044.

283. Postma, N. D., R. VanGiessel, and F. Reinink. "The Stirling Engine for Passenger Car Application," presented at the SAE Combined Commercial Vehicle Engineering & Operations and Powerplant Meetings, Chicago, Ill., June 1973, Paper No. 730648.

284. Beale, W., W. Holmes, S. Lewis, and E. Cheng. "Free-Piston Stirling Engines—A Progress Report," presented at the SAE Combined Commercial Vehicle Engineering & Powerplant Meetings, Chicago, Ill., June 1973, Paper No. 730647.

285. vanBeukering, H. C. J., and H. Fokker. "Present State-of-the-Art of the Philips Stirling Engine," presented at the SAE Combined Commercial Vehicle Engineering & Operations and Powerplant Meetings, Chicago, Ill., June 1973, Paper No. 730646.

286. Agarwal, P. D. "Electric Car and Air Pollution," presented at the SAE Automotive Engineering Congress, Detroit, Mich., January 1971, Paper No. 710190.

Appendix A

CONVENTIONS USED TO EXPRESS CONCENTRATION

VOLUMETRIC CONCENTRATION

The concentration of components in a multicomponent gaseous mixture can be expressed in a variety of ways, all of which reflect the relative number of molecules of a particular component as compared to the total number of molecules that comprise the mixture. The mol fraction and mol per cent are probably the most familiar to the reader. A volumetric unit that is often more convenient for expressing the concentration of trace species in a mixture is the part per million (ppm).

$$1 \text{ ppm} = \frac{1 \text{ volume of trace species}}{10^6 \text{ volumes of gaseous mixture}} \tag{A-1}$$

When trace substances are present in concentrations substantially less than a ppm, still smaller units such as the part per hundred million (pphm) or part per billion (ppb) may be convenient. These are defined as follows:

$$1 \text{ pphm} = \frac{1 \text{ volume of trace species}}{10^8 \text{ volumes of gaseous mixture}} \tag{A-2}$$

$$1 \text{ ppb} = \frac{1 \text{ volume of trace species}}{10^9 \text{ volumes of gaseous mixture}} \tag{A-3}$$

One of the advantages associated with these volumetric units is that they are all related by various powers of ten. These relationships are shown in Figure A-1. To convert A from the units given in the column

A \ B	mol fraction	mol %	ppm	pphm	ppb
mol fraction	1	10^2	10^6	10^8	10^9
mol %	10^{-2}	1	10^4	10^6	10^7
ppm	10^{-6}	10^{-4}	1	10^2	10^3
pphm	10^{-8}	10^{-6}	10^{-2}	1	10^1
ppb	10^{-9}	10^{-7}	10^{-3}	10^{-1}	1

Figure A-1. Relationship between common volumetric units of concentration.

on the left-hand side of the table to those in the row along the top of the figure multiply by the factor, f, in the table.

Algebraically,

$$B = fA \tag{A-4}$$

Another advantage of volumetric units is that so long as no condensation occurs they are independent of changes in temperature and pressure.

There are a few areas in which caution must be exercised. Equation A-3 and Figure A-1 are based on a billion being defined as a thousand million or 10^9. This is the definition used in the United States and France. In Great Britain and Germany a billion is defined to be a million million or 10^{12} and the conversion factors presented in Figure A-1 cannot be applied. Additionally, volumetric units refer to the ratio of gas volumes and not to ratio of weights or mass. These volumetric units cannot be used to represent the concentration of condensed phases (liquids or gases) suspended in a gas.

GRAVIMETRIC CONCENTRATION

Gravimetric concentration in a multicomponent mixture refers to the weight or mass of a species per unit volume of the mixture. Gravimetric units can be used to describe the concentration of gaseous components as well as of liquid and solid species suspended in a gaseous mixture. The most common gravimetric units encountered in air pollution and some source emission discussions are micrograms per cubic meter, abbreviated as $\mu g/m^3$ and milligram per cubic meter (mg/m^3).

Mixing of weight and volume units introduces a problem that does not exist with volumetric concentration. Since the mass of gaseous mixture in a cubic meter depends upon temperature and pressure, it is necessary to specify both these conditions if gravimetric concentrations obtained at different conditions are to be compared. Commonly used sets of standard conditions (STP) are 760 mm Hg (1 atmosphere) pressure and either 0°C (32°F) or 20°C (67°F) temperature. To convert gravimetric units from one set of conditions to another the following relationship can be used.

$$\left(\frac{\mu g}{M^3}\right)_2 = \left(\frac{\mu g}{M^3}\right)_1 \left(\frac{P_2}{P_1}\right)\left(\frac{T_1}{T_2}\right) \tag{A-5}$$

where absolute units are used for pressure, P, and temperature, T.

The gravimetric and volumetric concentrations of gaseous species in a mixture can be related through the ideal gas law $PV = NRT$.

$$(ppm) = \frac{(\mu g/m^3)}{(MW)} \times \frac{RT}{P} \tag{A-6}$$

Here MW is the molecular weight of the species and absolute units are used for pressure and temperature. Figure A-2 is a nomograph developed from Equation (A-6) for conversion between volumetric and gravimetric units when the pressure is atmospheric and the temperature is 0°C (32°F). A straight edge that intersects the right-hand scale at the point representing the molecular weight of the trace species will also intersect the equivalent values of volumetric and gravimetric concentration. When the volumetric concentration is ten or less, the units of gravimetric concentration on the center axis will be ($\mu g/m^3$). For volumetric concentrations between 10 and 1000 ppm, the gravimetric concentration units will be mg/m^3.

Care should be exercised in the use of gravimetric units to express the concentration of classes of compounds, for example, hydrocarbons, oxides of nitrogen and of sulfur oxides. These classes contain a multiplicity of molecular species with differing molecular weights. Since gas

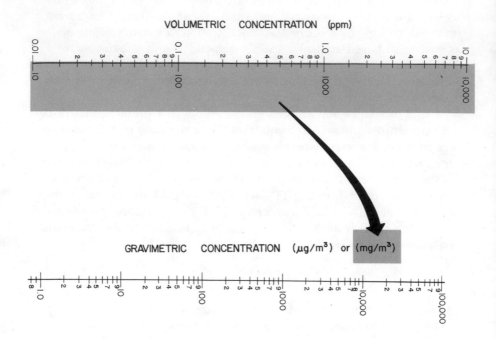

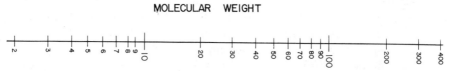

Figure A-2. Nomograph for the interconversion of volumetric and gravimetric units.

volumes are additive, the volumetric concentration for such a class is equal to the sum of the volumetric concentration of the components of that class. This is not the case for gravimetric concentrations. The usual convention is to express the gravimetric concentration of NO_x as if it were all NO_2 and of SO_x as SO_2. Hydrocarbons present problems in the cases of both gravimetric and volumetric concentration. Volumetric concentrations of gaseous species are often expressed in terms of ppm carbon. A molecule with four carbon atoms would then have a concentration of 4 ppm C. This is not a true volumetric concentration, however, and should not be treated as such. Gravimetric hydrocarbon concentrations are often expressed in terms of their equivalent methane (CH_4) value, that is, the weight of hydrocarbon per unit volume if it were all converted to methane.

ENUMERATIVE CONCENTRATION

To enumerate means to count. Consequently, an enumerative concentration refers to the number of particles contained in a unit volume of a gas or gas mixture. For example, sometimes particle concentration is reported as millions of particles per cubic foot (MMPCF). Enumerative and gravimetric concentrations are complementary. A gravimetric concentration conveys information about the weight of particles per unit volume of gas but nothing about the number of particles over which this weight is distributed. An enumerative concentration reveals the number of particles per unit volume but nothing about their mass.

Since the particles present in a gaseous mixture generally have a distribution of diameters, any meaningful statement of enumerative concentration must include upper and lower limits (a range) on the particle diameter included. Most particles both in the atmosphere and in the effluents of combustion sources are log normally distributed with respect to particle size. The value of enumerative concentration increases very rapidly as the lower limit of particle diameter decreases. Consequently an enumerative concentration is likely to overestimate the contribution of the smaller particles. On the other hand, as the particle diameter increases, the weight increases rapidly since it is proportional to the cube of the diameter and a gravimetric concentration overemphasizes the contribution of the larger and heavier particles.

Appendix B

CHEMICAL TERMINOLOGY

Historically, the subject of combustion has been more within the domain of the mechanical engineer than the chemist or the chemical engineer. Admittedly, individuals with chemistry backgrounds have been involved in researching the mechanisms by which combustion occurs. But, the statement is true within the context of engineering of the actual combustion devices and systems that emit pollutants to the atmosphere. The advent of concern about air pollution has led to the addition of chemical and pseudochemical terminology to the vocabularies of those who engineer combustion systems.

The objective of this appendix is to acquaint the reader with some of the terminology used in this book. Some of this terminology is derived from chemistry, for example, the names of specific chemical species. Other terms are pseudochemical: NO_x, a term used to refer to a mixture of nitric oxide and nitrogen dioxide in unspecified proportion, is an example. Such terminology may be abhorrent or confusing to the reader specifically schooled in chemistry. Consequently, a second objective of this appendix is to explain the rationale behind the evolution of such terms.

Combustion processes may give rise to literally hundreds, if not thousands, of different trace species that are emitted to the atmosphere. Our knowledge of this subject is limited by our ability to detect and measure the concentration of such emissions. Each year the advent of new and improved instrumental techniques reveals species hitherto undetected in the effluent of combustion systems.

Before proceeding further, it will be useful to distinguish between two classes of pollutants: primary and secondary pollutants. A primary pollutant is one emitted directly to the atmosphere by a source. A secondary pollutant is one arising as the result of chemical reactions of a primary pollutant with other species present in the atmosphere. In some cases, a single species may be produced by both mechanisms. For example, combustion processes produce two oxides of nitrogen, the oxide, NO, and the dioxide, NO_2. These are primary pollutants. The oxide, NO, subsequently undergoes oxidation in the atmosphere to form the dioxide, NO_2. In this sense, NO_2 is both a primary and a secondary pollutant.

Some of the more representative primary and a few of the secondary pollutants formed as a result of combustion will be considered.

The term *hydrocarbon* refers to an organic molecule that contains only hydrogen and carbon atoms. The simplest hydrocarbon molecule is methane, the formula of which is shown in the upper left-hand section of Figure B-1. The figures in this appendix show only the formulas

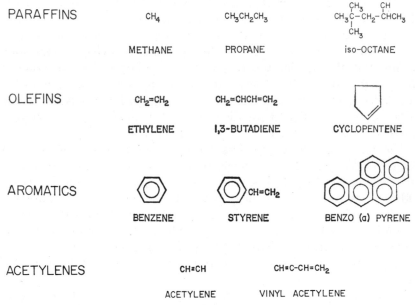

Figure B-1. Hydrocarbon species.

and, in some cases, a rather crude approximation of the actual molecular structure. In most cases, these molecules are three-dimensional, and the two-dimensional representations contained in this book are only a convenient shorthand. For a more complete description of these molecules, both in terms of their geometrical shape and their wave-mechanical interpretations, the reader is referred to an excellent text by Sebera.[1]

Methane is the first of a series of hydrocarbons referred to as paraffins or molecules containing carbon and hydrogen atoms connected with single bonds only. Each carbon atom is bonded to four other atoms (any combination of hydrogen and carbon atoms). Paraffins are saturated molecules in the sense that all of the available bonds are already present. Two other examples of paraffins are propane and isooctane. In the case of molecules containing more than three carbon atoms, it is necessary to distinguish between the case in which they are connected in a straight chain and the case in which they are connected in branched-chain fashion, as is the case for isooctane. The prefix *n-* or *normal* is used to denote a straight chain and, *i-* or *iso-* to denote a branched chain. The suffix *-ane* identifies paraffin species.

Olefins are species containing carbon atoms at least some of which are joined by double bonds. By contrast with the paraffins, which are saturated, the olefins are unsaturated. Additional bonds can be formed if a suitable species comes into contact with the double bond. It is this possibility that accounts for the relatively high propensity for unsaturated molecules to participate in atmospheric reactions and thereby produce secondary pollutants. The suffix *-ene* is used to denote any unsaturated species; *-diene* indicates two unsaturated bonds. Organic molecules with three or more carbon atoms, whether saturated or unsaturated, may be arranged in ring fashion. When this occurs, the prefix *cyclo-* is added to the name of the paraffinic or olefinic species. An example, cyclopentene, is shown in Figure B-1.

Aromatic hydrocarbons are compounds containing one or more six-membered carbon rings with each of the bonds connecting the carbon atoms in the ring being unsaturated. The simplest aromatic compound is benzene, which is denoted by an octagon with a circle circumscribed within it. There is a carbon atom at each of the apexes, and since each carbon atom can form four bonds, there is also a hydrogen atom connected to each carbon even though convention does not choose to show it. The prefix *cyclo-* is not used in the naming of aromatic molecules. Other examples of aromatic compounds are styrene and benzo-(a)-pyrene. Styrene is one of the possible products of combustion. It is of particular interest because it is a probable precursor for the formation of peroxy benzoyl nitrate, a secondary pollutant whose structure and properties will be considered shortly. Benzo-(a)-pyrene is an example of a polynuclear aromatic hydrocarbon, that is, one in which two or more aromatic rings contain common sides. These large and complex molecules are sometimes present in trace concentrations in liquid and solid fossil fuels. They may also be synthesized in pyrolytic reactions as illustrated in Figure 22. They are of particular concern because of their potential for carcinogenic activity.

The acetylenes are the final class of hydrocarbons that will be considered. These molecules contain one or more triple bonds between carbon atoms. The simplest is acetylene. Vinyl acetylene is another example of an acetylenic molecule. The acetylenes appear to be important intermediates in some combustion mechanisms, particularly those that are pyrolytic.

Of the hydrocarbons shown in Figure B-1, some like methane, propane and acetylene are gaseous at STP. Others such as isooctane and benzene are liquids and still others are solids, for example, benzo-(a)-pyrene. Thus, insofar as pollution of the air is concerned, the lower molecular weight hydrocarbons will be formed in the gas phase while the higher molecular weight ones may be present in condensed form. Hydrocarbons contain little or no electrical polarity and are therefore relatively insoluble in polar solvents such as water.

Figure B-2 presents the structures of a number of hydrocarbon

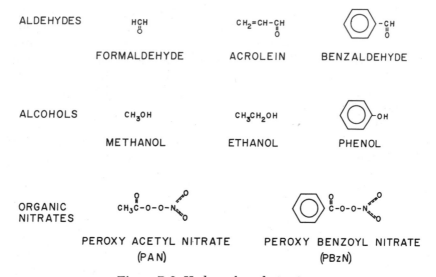

Figure B-2. Hydrocarbon derivatives.

derivatives. These molecules can no longer be referred to as hydrocarbons for they contain atoms other than hydrogen and carbon. They are derivatives of hydrocarbons in the sense that they can result from the reaction of hydrocarbons with other species either in the combustion process or later in the atmosphere. Three of the many possible classes are illustrated: the aldehydes, the alcohols and the organic nitrates.

Formaldehyde, a partial oxidation product of methane, is found in the effluent of some combustion processes. Acrolein and benzaldehyde are other examples of aldehydes. All contain the carbonyl group (−CH).
$$\overset{\|}{O}$$

The aldehydes may arise as both primary and secondary pollutants of combustion. They are of particular concern because of their lachrymatory properties.

Three alcohols are also shown. Methanol is the simplest, ethanol the next simplest, and phenol the simplest of the possible aromatic alcohols. All may arise as primary pollutants from combustion processes.

Two organic nitrates are shown: peroxy acetyl nitrate, or PAN, and peroxy benzoyl nitrate, or PBzN. Neither of these is a primary pollutant. Both are thermally unstable even at ambient temperatures and will not form at the higher temperatures encountered in combustion. Very low concentrations are formed as the result of the reactions of other primary pollutants in the atmosphere. Styrene, which was discussed earlier (refer to Figure B-1), is a probable precursor for the formation of PBzN, that is, styrene is likely to undergo reactions with species such as NO_2 and atmospheric oxidants to produce PBzN. Both of the organic nitrates shown are very potent lachrymatory substances.

Unlike the hydrocarbons, all of the aldehydes and alcohols shown are liquids. Furthermore, the presence of the alcohol group ($-OH$) induces a strong electrical polarity in the molecule, greatly enhancing its solubility in water. Thus, when these species are produced by combustion, they are likely to be found in the condensate unless precautions are taken to prevent this. The same is true in the atmosphere. They are likely to dissolve in any mist or other condensate that is present.

Figure B-3 shows the formula of four nitrogenous species. Normally,

N O NO_2

NITRIC OXIDE   NITROGEN DIOXIDE

Figure B-3. Nitrogenous species.

N_2O HNO_3

NITROUS OXIDE   NITRIC ACID

combustion processes emit nitric oxide and by comparison relatively small amounts of nitrogen dioxide to the atmosphere. The oxide is converted to the dioxide in the atmosphere. Nitrogen dioxide is an acid anhydride and readily hydrolyzes with the moisture of the atmosphere to form nitric acid, which is also shown. There are a number of other possible oxides of nitrogen, including the dimer of nitrogen dioxide, N_2O_4, and higher oxides such as N_2O_5 and N_2O_7. The concentrations of these oxides, both in combustion processes and in the atmosphere, is so low that they are not included in the aggregate term NO_x. NO_x is the sum of NO and NO_2. One other oxide, nitrogen suboxide with the formula N_2O, is also shown. It can be formed by some catalytically promoted reactions of NO in the effluent gases (refer to Section 3.4).

Figure B-4 shows four sulfur-containing species. Sulfur dioxide and

SO_2 SO_3

SULFUR DIOXIDE   SULFUR TRIOXIDE

Figure B-4. Sulfur-containing species.

COS H_2SO_4

CARBONYL SULFIDE   SULFURIC ACID

sulfur trioxide are both primary pollutants from the combustion of sulfur-bearing fuels. In addition, the SO_2 can be oxidized to SO_3 in the atmosphere. This is normally a slow reaction but since SO_3 is the thermodynamically stable species at STP it can be speeded considerably by the presence of an appropriate catalyst. The same is true for oxidation of SO_2 in the effluent gases, as was discussed in Sections 2.9 and 3.4. Sulfur trioxide is the acid anhydride of sulfuric acid, H_2SO_4. Carbonyl sulfide, COS, is another sulfur-containing molecule that under some conditions may form in the effluent gases of a combustion process. It is a potential product of the reaction of SO_2 and CO as discussed in Section 2.9.

This book focuses on the subject of pollution of the air by the emissions of trace species by combustion processes. The primary and secondary ambient air quality standards that have been established by the U.S. Federal Government are shown in Table B-I. Air quality standards have been established for six classes of species: SO_x (measured as SO_2), particulate matter, carbon monoxide, oxidant (measured as O_3), NO_x (measured as NO_2) and nonmethane hydrocarbons (measured as parts per million carbon). The potential importance of combustion processes as polluters of the air is apparent. With the exception of oxidants, all of these species are products of combustion. Furthermore, while oxidants are not direct products of combustion, they may be produced in the atmosphere by the reaction of other trace species that are products of combustion.

Of the classes of substances in Table B-I, all but one, carbon monoxide, are conglomerates, that is, they include more than a single molecular species. There are several reasons behind the evolution of these conglomerate classes. In the cases of particulates and oxidants, the most important reason is probably related to measurement technique. Particulates, for example, are normally collected on a filter. The volume of air passing through the filter is measured and the quantity of particulate collected is weighted to determine the gravimetric concentration. It is possible to analyze for the individual chemical species in the particulate matter; however, this is costly and time-consuming. Consequently, most of the data available on atmospheric concentrations of particulate substances and on the public health effects thereof are in terms of total suspended particulate matter. There is no doubt that some types of particulate matter are of much greater concern than others. At this time, our knowledge of the health effects of particulate substances at the levels encountered in ambient air does not warrant more specific standards. Note that at the higher concentrations encountered in industrial situations, individual Threshold Limit Values (TLV) have been established for different particulate substances (see footnote 15 of Table B-I).

The term *particulate matter* refers to solids or liquids present in a gas. Sometimes the term *aerosol* is substituted for *particles*. This classification presents a number of difficulties. First of all, from a strictly purist point of view, particles or aerosols may range from relatively large objects with characteristic diameters in the millimeter range to extremely minute statistical assemblages of molecules that may have only transient existence. Particles found in the atmosphere and in the effluents of combustion processes may vary in diameter by many orders of magnitude. Figure B-5 illustrates the range of diameter encountered in terms of examples, many of which will be familiar to the reader. This figure will be useful for comparing the diameters of particles formed during combustion (refer to Section 2.9) with universally familiar examples of particulate matter.

The size of a particle and the size distribution of a group of particles is an important property for classification purposes. The smaller and lighter particles remain suspended in the air, while the larger and heavier ones tend to settle out. These latter are referred to as dustfall in Figure B-5. A particle's size also influences its potential for producing adverse health effects. First of all, large particles are filtered out in the nasal region while the smaller ones may be carried deep into the respiratory system. Secondly, small particles contain large amounts of surface area per unit mass. Other species that may have greater

Table B-1
AMBIENT AIR STANDARDS[1]

Species	Averaging Period [2]	Air Quality Standards [3]		Episode Levels [4]			Industrial Health [5,6]
		Primary	Secondary	Alert	Warning	Emergency	TLV
SOx (as SO$_2$)	AAM	80(0.03)[7]	60(0.02)[7]				13,000(4.9)
	24	365(0.14), 1x[7]	260(0.09), 1x[7]	800(0.30)	1,000(0.38)	2,100(0.79)	
	8						
	3		1,300(0.49), 1x				
Particulate	AGM	75[8]	60[8]				—(15)
	24	260, 1x[8]	150, 1x[8]	375	750	1,000	
	8						
Carbon monoxide	8	10,000(9), 1x[9]	10,000(9), 1x[9]	17,000(15)	34,000(30)	46,000(40)	55,000(47)
	1	40,000(35), 1x[9]	40,000(35), 1x[9]				
Oxidant (as O$_3$)	8	160(0.08), 1x[10]	160(0.08), 1x[10]	200(0.1)	800(0.4)	1,200(0.6)	200(0.1)[14]
NOx (as NO$_2$)	AAM	100(0.05)[11]	100(0.05)[11]				9,000(4.7)
	24			282(0.15)	565(0.3)	750(0.4)	
	8						
	1						
Nonmethane Hydrocarbons (as CH$_4$)	3[13]	160(0.24), 1x[12]	160(0.24), 1x[12]	1,130(0.6)	2,260(1.2)	3,000(1.6)	—(16)
	8						

1 Format for each entry is as follows: STANDARD μg/m^3 @ 760 mmHg & 20°C (Equivalent Value, ppm). The maximum allowable exceedance rate, if any, follows. This refers to the maximum number of times per year that the standard may be exceeded. For example, 1× means the standard may be exceeded only once per year.

2 The averaging period is given in hours unless otherwise specified. AAM means Annual Arithmetic Mean Value and AGM means Annual Geometric Mean Value.

3 "National Primary and Secondary Ambient Air Quality Standards," *Federal Register* 36, #84, pp 8186–8201.

4 "Requirements for Preparation, Adoption and Submittal of Implementation Plans," *Federal Register*, 36, #158, pp 15486–15502.

5 "Occupational Safety and Health Standards; National Consensus Standards and Established Federal Standards," *Federal Register*, 36, #105, pp 10466–10714.

6 "Occupational Safety and Health Standards—Miscellaneous Amendments," *Federal Register*, 36, #157, pp 15101–15107.

7 As measured by Paraosaniline Reference Method. See Reference 3 above.

8 As measured by the High Volume Sampling Reference Method. See Reference 3 above.

9 As measured by Nondispersive InfraRed Spectroscopy Reference Method. See Reference 3 above.

10 As measured by the Chemiluminescence Reference Method for Ozone as described in Reference 3 above.

11 As measured by the Jacob-Hochheiser Reference Method for Oxides of Nitrogen. See Reference 3 above.

12 Total nonmethane hydrocarbons determined by FID. Reference Method described in Reference 3 above.

13 Three-hour average between 6 and 9 am.

14 TLV value for ozone not for photochemical oxidant.

15 No TLV given for total particulate. Individual TLV's for specific substances contained in particulate matter.

16 No TLV given for total hydrocarbons. Individual values specified for particular hydrocarbons.

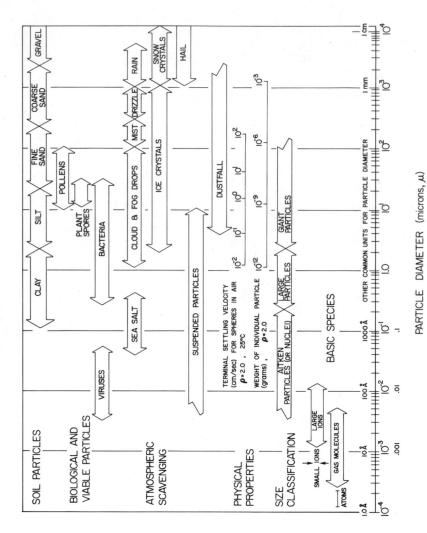

PARTICLE DIAMETER (microns, µ)

Figure B-5. Common particles.

biological activity than the particle itself can be adsorbed on this large surface area and carried deep into the respiratory system.

Another classification scheme that is of utility is the chemical composition of particulate matter. It is seen from Figure B-5 that particles may be composed of soil, of biologically active substances such as viruses and bacteria, and of other substances such as salt. The particles produced by combustion may contain carbonaceous substances, high molecular weight hydrocarbons such as benzo-(a)-pyrene and absorbed polar hydrocarbon derivatives. Collision and coalescence in the atmosphere of particles from different sources (*e.g.*, natural origin and combustion processes) may produce composite particles containing all of these different kinds of species.

Another conglomerate standard is that which exists for hydrocarbons. All hydrocarbon substances except the simplest, methane, are lumped together in a single standard. Methane is not included because it reacts much more slowly in the atmosphere than do the other hydrocarbons. Consequently, it does not contribute to the formation of secondary pollutants such as the oxidants.

NO_x denotes a mixture of NO and NO_2. The predominant product of combustion is NO, with only a small amount of NO_2 usually being formed. NO_2 is a thermodynamically stable species at ambient temperature, and the NO is oxidized to NO_2 in the atmosphere. Since one molecule of NO is oxidized to form one molecule of NO_2, the volumetric concentration of NO_x is equal to the sum of the volumetric concentrations of NO and NO_2. Furthermore, since these species have different molecular weights, the same is not true for gravimetric concentration. The NO_x standards are expressed in terms of the gravimetric concentration of NO_2. There are two reasons for this, the first being that it is NO_2 rather than NO which produces adverse health effects in the concentration ranges usually encountered in the atmosphere. The second reason is that the NO emitted is ultimately converted to NO_2 in the atmosphere, as discussed above.

The SO_x is expressed in terms of SO_2 rather than SO_3. SO_3 like NO_2 is a thermodynamically stable species at ambient temperatures. SO_3, however, is a very hygroscopic substance and will react with any moisture in the atmosphere to form an acid mist (sulfuric acid). The mist is a condensate rather than a gas and will form even when the relative humidity is far below saturation. Under normal conditions (*i.e.*, in the absence of a catalyst) the atmospheric oxidation of SO_2 to SO_3 is rather slow. As a result, the ambient air quality standard for SO_x is expressed in terms of the gaseous species SO_2 rather than SO_3, which is a condensate and may be included in particulate measurements.

Evidence is accumulating that the simultaneous presence of SO_x and particulate matter has adverse effects on public health, effects greater than the effects of either one above. The earlier discussion of how particulates may absorb other substances and carry them deep into the respiratory system is an example of how this might occur. This phenomenon has been referred to as synergy. Understanding in this area is still in the early stages of development. It may be that the SO_2 and SO_3 are converted to sulfate ion in the particulate matter, and it is this sulfate or some fraction of it, perhaps the soluble fraction, that is responsible for adverse health effects. If this is true, an air quality standard for suspended sulfate may be desirable either in place of or in addition to the standards in Table B-I.

Finally, oxidants as noted above are not direct products of combustion processes but arise from atmospheric reactions of other species, some of which are produced during combustion, notably NO_2 and the nonmethane hydrocarbons. Chemiluminescent measurement of atmospheric ozone is used as the index of oxidants. Ozone was chosen as the index partly because of its adverse health effects and partly because of the availability of a reliable reference method for measuring it. A number of other substances are also oxidants, for example, the PAN and PBzN molecules shown in Figure B-2. The use of ozone as an index of ozone is not entirely satisfactory. The eye-irritating characteristic of some types of air pollution is usually associated with oxidant. Ozone is a product of atmospheric photochemical reactions; consequently high concentrations are not formed on overcast days. Yet on these same overcast days, significant eye irritation may be experienced. Therefore, the evaluation of new and more specific standards, perhaps based on measurements of a substance such as PAN, seems a desirable goal.

Table B-I contains a host of other information that will be addressed briefly here. Air quality standards are established for different averaging times ranging from an hour to a year. The shorter averaging times of the primary standards are intended to limit the public to exposures of short term high concentrations that may produce adverse health effects. The longest averaging period, a year, can be useful as a measure of the general trend in the air pollution of an area. The absence of a short term standard does not mean that short term high concentrations have no adverse effects. In the case of NO_x the only air quality standard that exists is the annual arithmetic mean. It is generally agreed that there is a need for a shorter term standard, perhaps a 24 hour one; however, the available data at the time the standards were set was inadequate to legally support such a standard. Even today, there is considerable confusion in this area. The Jacob-Hochheiser reference measuring technique (see footnote 11 of Table B-1) has been found to be in error and a new reference method will have to be proposed. Reevaluation of the information on ambient concentrations of NO_2 and of the health effects of the same must follow.

The threshold limit values are based on 8-hour averages because this is the length of a typical working day. The 6-9 a.m. three-hour averaging period for hydrocarbons (see footnote 13 of Table B-I) is chosen because it is believed that the hydrocarbon emissions during this period of time are subsequently radiated by the ultraviolet component of sunlight and result in the formation of high oxidant and NO_2 levels later in the day.

The distinction between primary and secondary air quality standards is as follows. Primary standards are intended to protect the public health, whereas secondary ones may be more stringent and are intended to protect the public welfare. Public welfare includes a broad range of considerations including the effects of air pollutants on water, soil, crops, other vegetation, domestic animals and wildlife, atmospheric visibility, weather, climate, economic values and on personal comfort and well-being.

Atmospheric pollution levels that are used to define the existence of an air pollution episode are also shown in Table B-I. The term *episode* is rather self-explanatory. It refers to incidents that are usually infrequent and during which abnormally high levels of atmospheric contamination occur. These high contamination levels are caused by abnormal meteorological conditions that severely restrict the atmospheric dispersion of pollutants. Various courses of action are suggested or required by law as the severity of an episode increases. These range from recommendations that persons with respiratory problems remain indoors and not engage in strenuous activities to court injunctions that shut down all but very essential activities during a severe episode.

The discussion in this appendix has digressed considerably from the subject matter of this book, *i.e.*, combustion. It is a worthwhile digression because ultimately it is considerations such as these, public health and public welfare, that determine whether the trace species emitted by combustion processes are of concern or not.

REFERENCE

1. Sebera, D. K. *Electronic Structure and Chemical Bonding* (New York: Blaisdell Publishing Co., 1964).

EQUATION INDEX

SUBJECT INDEX

229